BLUEBIRD REBORN

LCC No. 1 emerging from Kingsway Subway in London during a Southern Counties Touring Society tour in 1948. (John Wootton Collection/National Tramway Museum)

BLUEBIRD REBORN

The History and Restoration of LCC No. 1

The right to be identified as the Authors has been asserted in accordance with sections 77 and 78 of the Copyright Designs and Patent Acts 1988.

All rights reserved. No part of this publication may be reproduced, stored in a retrieval system or transmitted, in any form by any means, electronic, mechanical, photocopying, recording or otherwise, without prior permission in writing from the publisher.

Published by: The Light Rail Transit Association
8 Berwick Place
Welwyn Garden City AL7 4TU

www.lrta.org

Copyright © Light Rail Transit Association 2023

ISBN Number: 978-0-948106-69-9

Printed by: Page Bros Group Ltd
Mile Cross Lane
Norwich NR6 6SA

www.pagebros.co.uk

Front cover:
LCC No. 1 on the Depot fan in March 2023. (National Tramway Museum)

Rear cover top:
Some of the many volunteers and staff who worked to restore LCC No. 1 at the National Tramway Museum, seen on the Depot fan in April 2022. (J Dignan)

Rear cover bottom:
No. 1 operating on Service 35A in 1930s London.
(M J O'Connor/National Tramway Museum)

Contents

Foreword 6

Preface 7

Chapter 1
Bluebird: a symbol of hope in turbulent times? 9
Assessing the prospects for a tramway revival during the late 1920s 11

Chapter 2
Revival hopes derailed: challenging location and unfortunate timing 19
Challenging location: the public transport context in London 19
Unfortunate timing for LCC No. 1's inception: the makings of a 'perfect storm' 21

Chapter 3
Revival hopes frustrated: underlying obstacles 26
Operational and regulatory constraints 26
Municipal parochialism and obstructionism 28
Failure to reform the regulatory framework 32
Free-for-all for London tramways' rivals 36
Consolidation and empire-building by London tramways' rivals 39
LCC and its tramway: hapless victims or hopeless strategists? 41

Chapter 4
The need for London County Council No. 1 45
Changes and limitations 45
Procedures and personnel within the LCCT 45
New tramcars to meet traffic demand 45
The bus as the major competitor 46
Developments of the motor bus 47
The LCC's position on tramways in 1924 49
Forward thinking by the management of LCC Tramways 50
In the 1920s, was the trolleybus a competitor? 52
Underground competition and extensions 55
Modernisation of existing LCC tramcars 55
Fares and publicity 57
Modernisation developments by the London tramway companies with new tramcars 57
LCC Tramways' replacement programme and modernisation with new tramcars 63
More LCC tramcar orders 65
Developments in tramcar bodywork construction technology 66

Chapter 5
Experimental Tramcar 71
Design 71
Construction 74
Associated experimental work 78
The media reception 80
LCC No. 1's influence on tramcar design 82

Chapter 6
No. 1 in London 87

Chapter 7
LCC No. 1 becomes Leeds City Transport 301 102

Chapter 8
Preservation 111
The British Transport Commission and its Consultative Panel 111
The Museum of British Transport, Clapham 112
LCC No. 1 comes to Crich 115
Stabilisation and conservation 115
Preparations for restoration: the 2012 Condition Assessment 119

Chapter 9
Restoration 123
An Offer for Restoration 123
Funding 124
Scheduling a move to the Conservation Workshop 126
The Restoration Programme 127
Trucks and associated electrical work 127
Long lead and missing items 128
Labour resource 128
Saturday 14th June - the restoration begins 129
Restoration 129
Bodywork and associated mechanical work 130
Trucks and associated electrical work 157
Air supply system 167
Electrical supply system and associated mechanical equipment 172
Testing and commissioning 180
And finally 184

Appendix 1
The Tramway and Railway World and *The Electric Railway, Bus and Tram Journal* coverage of LCC No. 1, May 1932 186

Appendix 2
Hidden Stories 191

References 199

Acknowledgements 209

Author Biographies 210

General Index 211

Index for LCC No. 1 (Bluebird) 221

Foreword

Ian Ross

President, The Tramway Museum Society (2020-2022)

Chairman, London County Council Tramways Trust (2002-present)

It is a privilege to be asked to write a foreword to this history of Bluebird, London County Council Tramways' magnificent prototype tramcar No. 1. My personal involvement in the preservation of historic tramcars in the collection of the National Tramway Museum runs almost concurrently with No. 1 from its arrival in December 1972 to the completion of its restoration. When I helped in a small way with its removal from the Museum of British Transport at Clapham I could not have foreseen that I would go on to be actively involved in both raising funds for restoration and in the restoration process itself.

The history of No. 1 is inextricably linked with factors – economic, political and social – that affected public transport in London, and indeed throughout the United Kingdom, after 1930. The survival of this iconic tramcar into the era of preservation was also due to similar factors in Leeds after the Second World War. The management and engineering teams within the London County Council Tramways deserve significant praise for producing a prototype tramcar that aimed to arrest the decline in passenger numbers brought about by the growth of private motoring and improvements in passenger comfort in other forms of public transport in London. It is a matter of regret that changes in transport policy after 1933 meant that Bluebird was destined to be a solitary example of 'what might have been'.

Research undertaken by the co-authors of this book, and by the engineering team at the National Tramway Museum, has confirmed much of what has been known previously about LCC No. 1. However a significant amount of new information has come to light, some of which casts doubt on earlier histories. Records of the tramcar disposal plans drawn up by London Transport have been made available from original documents that were not known to have survived; these show that LCC No. 1 was expected to remain in service until the very end of London's tramways. Engineering investigations during deconstruction of the tramcar's bogies showed that material shortages were not unknown in the nineteen-thirties leading to substitution of inferior components. Many other examples of revisions to Bluebird's story are recorded in this book.

It is often customary for the author of the foreword to any historical record to acknowledge the contribution made by individuals and organisations to the final work. I have given this aspect deep and careful thought and I have decided to keep the list of named persons short and, I believe, particularly relevant to the history of LCC No. 1. In doing so I regard the contribution of the engineering and curatorial teams of the National Tramway Museum equally highly; without them this magnificent restoration would not have been achieved. I must also record my grateful thanks to the supporters of the London County Council Tramways Trust for their generous financial support over a long period. For securing Bluebird's survival I commend John Scholes of the British Transport Commission, Richard Elliott and Victor Matterface of London Transport, Leeds City Council Transport Department and John Price who was a member of the BTC Consultative Panel on Historical Relics. Finally if I may be allowed an amount of professional bias as an engineer I acknowledge the work of Messrs Ireland and Sinclair, Rolling Stock Engineers of the LCC Tramways, and their leading designers Messrs Harding and Rivett. Without their expertise Bluebird would not have seen the light of day.

Crich, Derbyshire
March 2023

Preface

Laura Waters
Curator Collections & Library, The National Tramway Museum (2013-2021)

Kate Watts
Curator, The National Tramway Museum (2021-present)

In May 2017, I first met with my co-authors and copy editor to discuss this book. My task then, as Curator for the National Tramway Museum, was to coordinate a team who could deliver the Tramway Museum Society's 2012 Board of Management resolution to create "explanatory and interpretation media" which would do justice to the deconstruction, thorough investigation and restoration of this stunning 1930s tramcar, London County Council No. 1 – Bluebird.

We could have settled for the easy task of producing a publication or exhibition chronicling the restoration of the vehicle. This, for many, would have provided a fascinating insight into the practices of the National Tramway Museum's Conservation Workshop, who work tirelessly to complete world class restoration projects, whilst maintaining the Museum's Nationally Designated tramcar collection for continued demonstration to the public.

However, in 2017 the Museum's Curatorial Department was at a developmental point in our five year plan, and we were undertaking fresh research into our collections. With LCC No. 1 we had, for the first time in a number of years, the opportunity, knowledge and resources to deliver a more thorough, academic piece of research that would link firmly to the Museum's core purposes and "share, inspire and educate through telling the story of tramways".

We would examine the tramcar's recent restoration, and delve further to explore the more than 90 years' worth of history it represents; we would reveal stories demonstrating its place in both the history of London's tramways and the development of the British tramcar as a whole.

When the book team came together, the Museum's Conservation Workshop team were in their third year of work on the tramcar. Already there were new stories to explore, as well as evidence which challenged accepted information about the vehicle.

We decided that for the book to read as a comprehensive history, we would split it into two main sections.

The first section includes chapters discussing the political, social and economic background to the development of LCC No. 1, and the hopes of London County Council Tramways that the tramcar might revive their fortunes. Further chapters detail the design and construction challenges of the tramcar, which have in many ways been mirrored in our Conservation Workshop team's endeavours to restore it. The final chapters in this section tell the tramcar's story from service in London to a new home and further operation in Leeds, then through preservation and acknowledgment of its position in tramway history. The section concludes with the tramcar in preservation at the National Tramway Museum.

The second section focuses on the restoration of LCC No. 1. It explains not only how the Museum decides whether or not to restore a tramcar within its collection, but also the process that leads up to the physical restoration taking place. Supported by images taken throughout the duration of the project, this section takes the reader through every major element of the restoration of London County Council No. 1 to 1932 operational condition.

Whilst researching and writing this book, my co-authors and I have sought to bring together as much original research and source material as possible. The Museum's own subject specialist Library & Archive has proved as ever to be a rich source of information, with historic journal articles, books, committee minutes, photographs, letters and Acts of Parliament all consulted as part of our research. We have also looked to other, equally rich sources of information: London Metropolitan Archives, the home of the London County Council Archive; the collections of London Transport Museum and the London County Council Tramways Trust; and individuals who have kindly shared their experience of LCC No.1, and their private collections.

Each individual author has followed their own avenues of research for their respective chapters, and their unique voices come through in each one. In addition, during the project they have supported one another, shared knowledge and new sources of information, and challenged each others' assumptions, and it is this which brings the publication together, not merely as a collection of standalone chapters, but as a whole.

Our journey could not have evolved as it has and culminated in this book without our copy editor, who has shaped and moulded the text, checked our use of language to explain facts, figures and technical details, and taught all of us authors many a thing about the consistency of writing.

Although I handed the National Tramway Museum Curator's position over to Kate Watts in 2021, and, with it, oversight of the completed restoration of LCC No. 1 and the development of this book, I look back with pride on the vast amount of work which has been undertaken to bring the book to fruition, and how far we have come from our early drafts.

Laura Waters

Having arrived to take the post of Curator at the National Tramway Museum during the end stages of this project, I cannot claim to have made any significant contribution to it. However, having observed the final push to complete this complex tramcar restoration, I have been drawn into the fascinating story of LCC No. 1. From design and production through to later working life, being saved for preservation and finally reborn at Crich, the story of Bluebird deserves to be recorded. This book is the result of years of research by staff and volunteers alongside the physical restoration project. We hope it succeeds in telling the story of LCC No. 1 whilst also giving an insight into how the National Tramway Museum utilises the skills and expertise of its team to carry out professional first-class tramcar restorations.

Kate Watts

Chapter 1
Bluebird: a symbol of hope in turbulent times?
Jim Dignan

On 24 October 1929, approval was given by London County Council's Highways Committee for the construction of a new experimental tramcar with a view to trialling new methods of construction and equipment [1]. The tramcar was subsequently launched amid much publicity on 5 May 1932 and was formally designated as LCC No. 1. It entered service one month later, on 10 June 1932, and was enthusiastically welcomed by press and public alike. Press reviews heralded "A tramcar revolution" and spoke of "Rolls-Royce" levels of passenger comfort [2].

This was not an auspicious time for the tramways industry, however. A quarter of a century since the start of the electric tramway era, many of the tramcars were in acute need of renewal or refurbishment, as was much of the tramway infrastructure. On many of Britain's tramways the overall condition of tramcars, track and overhead equipment had deteriorated markedly during the First World War. Economic disruption continued even after the war was over, which hampered investment and the task of reconstruction. The enduring period of austerity also contributed to an increase in industrial unrest, as soldiers who returned to peacetime occupations found that their expectations of a better life were not being met.

The ending of hostilities a decade earlier had also resulted in a sudden decommissioning of hundreds of buses that had been used as troop carriers. Many of these were snapped up and returned to civilian passenger service by private operators in competition with existing tramways. Throughout the 1920s, ageing tramcars also faced increasing competition from more modern designs of motor buses. These

Painting by Ashley Best showing LCC No. 1 emerging from the northern ramp of the Kingsway Subway in 1932. This part of the subway survives and has changed little over the last eighty years.
(LCCTT Collection)

benefited from the introduction of pneumatic tyres and other advances in passenger comfort, together with dramatic improvements in petrol engine technology, as experience gained during the Great War resulted in increased power and reliability.

The post-war era also coincided with the dawn of mass motor vehicle production, in factories such as the Ford plant at Trafford Park, which contributed to a dramatic increase in the number of private cars that began competing in earnest with trams for available road space.

In the face of these mounting challenges, a growing number of tramway systems – 60 mostly smaller operators across the United Kingdom, including 39 in England alone – had closed down altogether during the 1920s, and switched instead to fleets of motor buses or trolleybuses.

Other tramway operators, including London County Council, adopted a different strategy in response to the growing competition they faced, pinning their hopes on the development of more modern tramcars offering greater levels of passenger comfort and improvements in operational efficiency. LCC No. 1 was not the first attempt by the LCC to upgrade and modernise its fleet, as will be seen in chapter 4, but it was the most radical and ambitious to date.

One of the most distinctive features of the new tramcar was its striking royal blue and ivory livery which accentuated its streamlined appearance and set it apart from the rest of the LCC fleet. Not surprisingly, perhaps, the tramcar swiftly acquired the nickname 'Bluebird', and was referred to as such by passengers and tram crew alike, one of very few individual tramcars to be acknowledged in this way.

The choice of nickname could also have been influenced by the prominence that was given in the news outlets at the time to the exploits of Malcolm Campbell in his high-profile pursuit of the world land speed record in a car that he had also christened 'Blue Bird'. Indeed, cars bearing the same name and driven by Campbell broke the land speed record on no fewer than six subsequent occasions between 4 February 1927 and 3 September 1935, and Campbell himself was knighted in February 1931, the year before LCC No. 1 was launched.

Campbell's inspiration for adopting the name 'Blue Bird' - and also the striking azure blue colour scheme with which it was associated - was an eponymous play written in 1908 by the Belgian playwright Maurice Maeterlinck, in which the blue bird features as a symbol of happiness. Immediately after seeing the play, Campbell is said to have been inspired to rename his car and to have it repainted overnight [3]. The choice of colour scheme for LCC No. 1 seems to have been a more considered affair as a 1" scale model of the tramcar was produced at the Charlton Works so that various styles of livery could be tried out for the approval of the Tramways Department [4].

With its ground-breaking design and advanced specification, LCC No. 1 was clearly envisaged as a precursor for a new class of tramcars, intended to augment the LCC's ambitious modernisation programme, which was already well under way. The fact that the experimental tramcar was designated No. 1 and that blank fleet numbers 2-100 were available led both J H Price and J Reed to surmise that a class of 100 'Bluebird' derivatives might have been envisaged [5] [6].

Others claimed that the figure might even have been higher. E R Oakley, for example, referring to the LCC's ongoing tramcar modernisation programme, stated: "…it was intended that the second batch of 150 new cars that were being considered at that time would also be built to the new [LCC No. 1] design." [7]

In January 1927, Major (later Sir) Malcolm Campbell in his famous 'Blue Bird' broke the world's Land Speed Record at Pendine Sands. (www.commons.wikimedia.org)

Despite these hopes and expectations, Bluebird was destined to remain a 'one-off'. Just a year after its launch, the London County Council's Tramways Department was itself subsumed into the newly created London Passenger Transport Board (LPTB) as part of a major reorganisation of London's public transport. Within five years the tramcar itself had lost its distinctive livery and the days of all tramcars in London were themselves numbered following the LPTB's adoption of a policy of tramway abandonment.

So, what went wrong? Did Bluebird fail to live up to the hopes and expectations that were placed upon it? Was it realistic to imagine that a redesigned fleet of tramcars could act as a catalyst for a tramway revival during the inter-war period? If so, would such a revival have been feasible in a London tramway context? Did the LCC and its Tramways Department fail to mount an effective response to the various challenges they faced? Or were both tramcar and tramway hapless victims of forces and circumstances that they were powerless to confront or control?

Assessing the prospects for a tramway revival during the late 1920s

In spite of the formidable challenges that Britain's remaining tramway operators faced, there were realistic grounds – particularly in some of the larger cities – for hoping that the situation might still be retrievable. Local authorities had been one of the key drivers behind the initial development and expansion of the tramway network around the start of the 20th Century. One ground for optimism lay in their continuing potential to maintain their support for, and investment in, their local tramways.

Municipal tramways were capable of providing important reciprocal benefits to the local economy, such as fares revenue and employment opportunities. Several larger operators also developed significant tramcar manufacturing capabilities, which further enhanced their economic significance. In addition, a number of councils, including the LCC, made use of their tramways to support other social policy and urban planning initiatives, for example in the field of housing and slum clearance.

In line with the then prevalent doctrine of municipalisation, local authorities enjoyed wide-ranging autonomy and also revenue-raising powers that enabled them to own and operate a wide range of utilities – including gas, water and electricity as well as public transport – in order to provide essential services for their residents safely, fairly, accountably and at reasonable prices [8].

Many local authorities across the political spectrum continued to avail themselves of these powers and in a number of instances around the country (and also abroad) chose to invest in, renew and develop their municipal tramways even in the face of competitive challenges that were similar to those confronting the LCC in the late 1920s.

Sheffield 61, one of the more modern tramcars built at the Corporation's Queens Road works in 1930. (M J O'Connor/National Tramway Museum)

Leeds Middleton Bogie 255 on reserved track at Easterby End, 1952.
(R B Parr/National Tramway Museum)

Sheffield Corporation, for example, embarked on an ambitious modernisation programme in 1929, the very year that LCC No. 1 was commissioned, to replace its entire fleet of life-expired early electric tramcars, most of which had previously been purchased from specialist suppliers, with a fleet of new vehicles, the majority of which were constructed in the Corporation's own workshops. Over the next decade 213 new tramcars were added to the fleet, which helped to extend the life of the tramway until its eventual closure in October 1960 [9].

Leeds was another city that adopted an extensive modernisation and refurbishment programme that commenced in 1930 with the introduction of over 100 trams that were built to a modern design, which paved the way for a succession of different designs in subsequent years [10]. Leeds was also an early pioneer in the use of tramway extensions using reserved tracks to serve new housing developments such as the one at Middleton on the outskirts of the city, which was completed during the 1920s. Indeed, an entirely new class of tram, the Middleton Bogie, was designed and built specifically to service this route [11]. As with Sheffield, these investments in rolling stock and infrastructure enabled the tramway to survive until well after the Second World War before it eventually closed in 1959.

By the late 1920s Sunderland's trams and infrastructure were largely life expired and the service on one route (Villette Road, a service restricted to single deck tramcars on account of a low overhead railway bridge) was withdrawn for a time in favour of buses in August 1930 [12]. However, Charles Hopkins, the recently appointed manager, designed and commissioned a new experimental single-deck tramcar with low entrance steps, comfortable seating and improved performance. Its successful introduction not only prompted a return of trams on the Villette Road route but also relaying of the track across the entire tramway combined with a major investment in new, second-hand and refurbished rolling stock that extended the life of the tramway to 1954.

At around the same time, in 1932, Liverpool Corporation, the third biggest tramway operator in England at the time, also embarked on a massive tram building and refurbishment programme that resulted in the construction of nearly 300 modern tramcars of various designs over the next decade [13]. Liverpool had also pioneered the development and use of central reservations – so-called 'grass tracks' – in 1914 that speeded up tramcar services and were incorporated in the numerous tramway extensions that were opened during the inter-war period. Many of these were designed to serve a series of new suburban housing estates which enabled the Corporation to move thousands of residents from densely-packed inner-city 'slums'. Liverpool's tramway continued in operation until the years of war-time neglect contributed to its eventual closure in 1957.

Sunderland experimental single deck tramcar 85 at Hylton Road Depot Yard, date unknown. (H Nicol/National Tramway Museum)

Yet another major British city, Glasgow, with one of the largest tramway systems in Europe was unusual in being one of the first operators to open its own tramcar workshop, which was heavily involved in the construction of the 1,000-strong tramcar fleet almost from the outset. Another innovation was its early decision to institute an ongoing programme of refurbishment and upgrading of its entire fleet on a regular basis as an alternative to one-off replacement programmes. It did, however, also construct two new prototype tramcars during the mid-1920s, which formed the basis of a new class of 50 Kilmarnock Bogie tramcars that were delivered between 1927 and 1929 [14]. Other new designs were introduced during the 1930s, and the tramway outlasted all other British city tramways before finally closing in 1962.

Looking further afield, many continental European cities also successfully extended the life of their tramways with the aid of modern tramcars and investment in infrastructure during the inter-war era.

Liverpool Streamliner 950 on reserved tracks at Fazakerley. (R B Parr/National Tramway Museum)

One example among many that is of particular interest in the present context is the city of Milan, which combined a radical restructuring of its route network in 1926 with the acquisition of 500 new tramcars [15]. These were modelled on a Peter Witt design that had been adopted in Cleveland among other US cities. Following the construction of two successful prototypes that were designed by Milan's municipal engineers in 1927, orders for 500 of the resulting ATM Class 1500 tramcars (also known as Type 1928) were placed with a number of local manufacturers and deliveries took place between 1929 and 1930. Despite the subsequent introduction of trolleybuses on some peripheral routes, not only has the Milan tramway itself survived to this day, but a significant number of the Class 1500 tramcars were still operating more than nine decades after their initial introduction [16].

Another equally significant development that also took place in 1929, the year LCC No. 1 was commissioned, was the establishment in North America of a design committee known as the Presidents' Conference Committee (PCC) that comprised representatives of a number of US streetcar operators and potential manufacturers [17]. Tasked with the challenge of designing a streetcar that would be capable of fending off growing competition not only from diesel buses but also the automobile, the committee came up with a radical new design of streetcar that was streamlined, comfortable, quiet and combined rapid acceleration with powerful braking to help reduce journey times. Costs were reduced by incorporating a high degree of standardisation that facilitated the adoption of mass production techniques.

The design was licensed to various constructors in North America and also in a variety of mainland European countries during the second half of the 20th Century. Just under 5,000 PCC streetcar units were produced in the USA between 1936 and 1952, plus many thousands more units in Europe, making it one of the most successful tramcar designs of all time. A major reason for the PCC cars' success was that operators enjoyed not only lower operating costs but also increased ridership. Of the many operators that adopted the PCC, the majority survived into the post-war period and although most of them subsequently abandoned the streetcar in favour of bus fleets, a number of operators survived into the 1980s. By that time the majority of PCC cars had been replaced by more modern light rail transit vehicles.

Not all tramways were municipally owned and operated, of course, though a majority of British ones were. Interestingly, a small number of privately run 'company' tramways also demonstrated during the inter-war period that closure in the face of competition from rival modes of transport was not inevitable if fleets were modernised and infrastructure maintained. One example was Gateshead & District Tramways Co., which formed part of the giant British Electric Traction (BET) conglomerate.

BET was for many years the largest privately-owned tramway conglomerate in the country, operating upwards of 40 separate tramway systems in the United Kingdom and abroad. Although most BET

ATM Class 1500 tramcars 1504 and 1709 in front of the old Central Station in Milan, 1929. (www.skyscrapercity.com)

tramways ran into financial difficulties and closed down during the inter-war period, Gateshead bucked the trend by modernising its fleet of tramcars between 1923 and 1928 and adopting a number of innovative policies to improve the efficiency of the fleet [18].

For example, faced with an awkwardly located low bridge close to Gateshead station at the hub of the network, which affected a number of routes, the company built a number of extra-long bogie-mounted single-deckers with far greater standing capacity than seated. In order to facilitate the boarding and alighting of such numbers, an innovative 'passenger flow' system was also adopted whereby passengers boarded the tram at the rear, using sliding doors that were operated by the conductor, and exited from the front using a hinged door that was controlled by the driver. Such practices enabled the Gateshead fleet to survive until August 1951, longer than any other BET operator.

In the light of these examples, in which the introduction of modern tramcar designs helped to stimulate a tramway renaissance in various cities at home and abroad, the idea that London might experience a similar revival was not, on the face of it, an altogether fanciful notion.

Indeed, there were some early indications that a successful fight-back might also be possible in the capital itself. For example, two privately owned and operated tramways, Metropolitan Electric Tramways

PCC-style streetcar at Connecticut Avenue Presbyterian Church, Washington DC, 1960.
(M J O'Connor/National Tramway Museum)

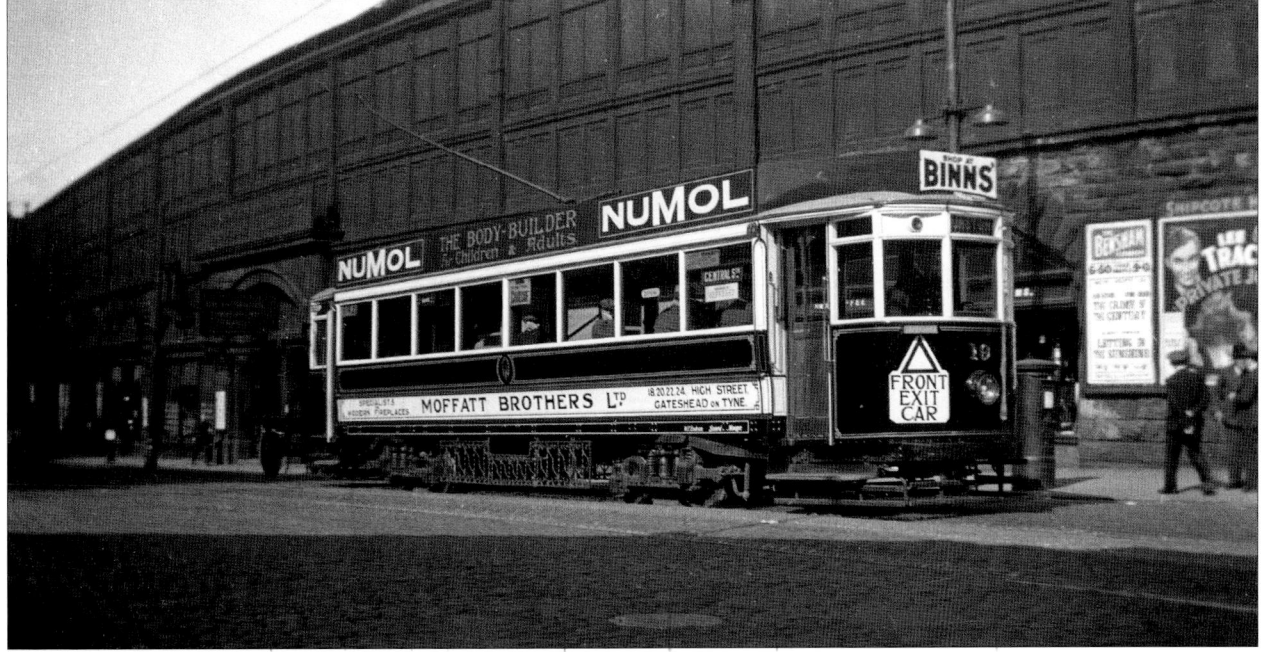

Gateshead single-deck bogie car 6 outside Gateshead station, date unknown.
(M J O'Connor/National Tramway Museum)

Metropolitan Electric Tramways experimental tramcar 318, also known as 'Bluebell', on 4 August 1929. (H Nicol/National Tramway Museum)

(MET) and London United Tramways (LUT), had each produced new experimental tramcars towards the end of 1927 that were followed by no fewer than three additional prototypes, each of which incorporated slightly different features. These will be described more fully in Chapter 4.

Experience gained from these prototypes helped to demonstrate that more modern designs of tramcar were capable of attracting additional passengers and generating fare revenue, in spite of the ongoing competition from buses. Indeed, they in turn formed the basis of an innovative and successful modern design, the Feltham tramcar, built by the Union Construction Company (UCC) based at Feltham, Middlesex (hence the name). 100 of these tramcars entered service with the two operators in 1931, 54 having been ordered by MET and 46 by LUT [19].

In addition to this investment in rolling stock, MET had relaid a substantial proportion of its tracks between 1929 and 1933. As a result of these various improvements, MET were running some of the fastest urban tramway services in Europe at an average speed of 12 mph with four stops per mile [20].

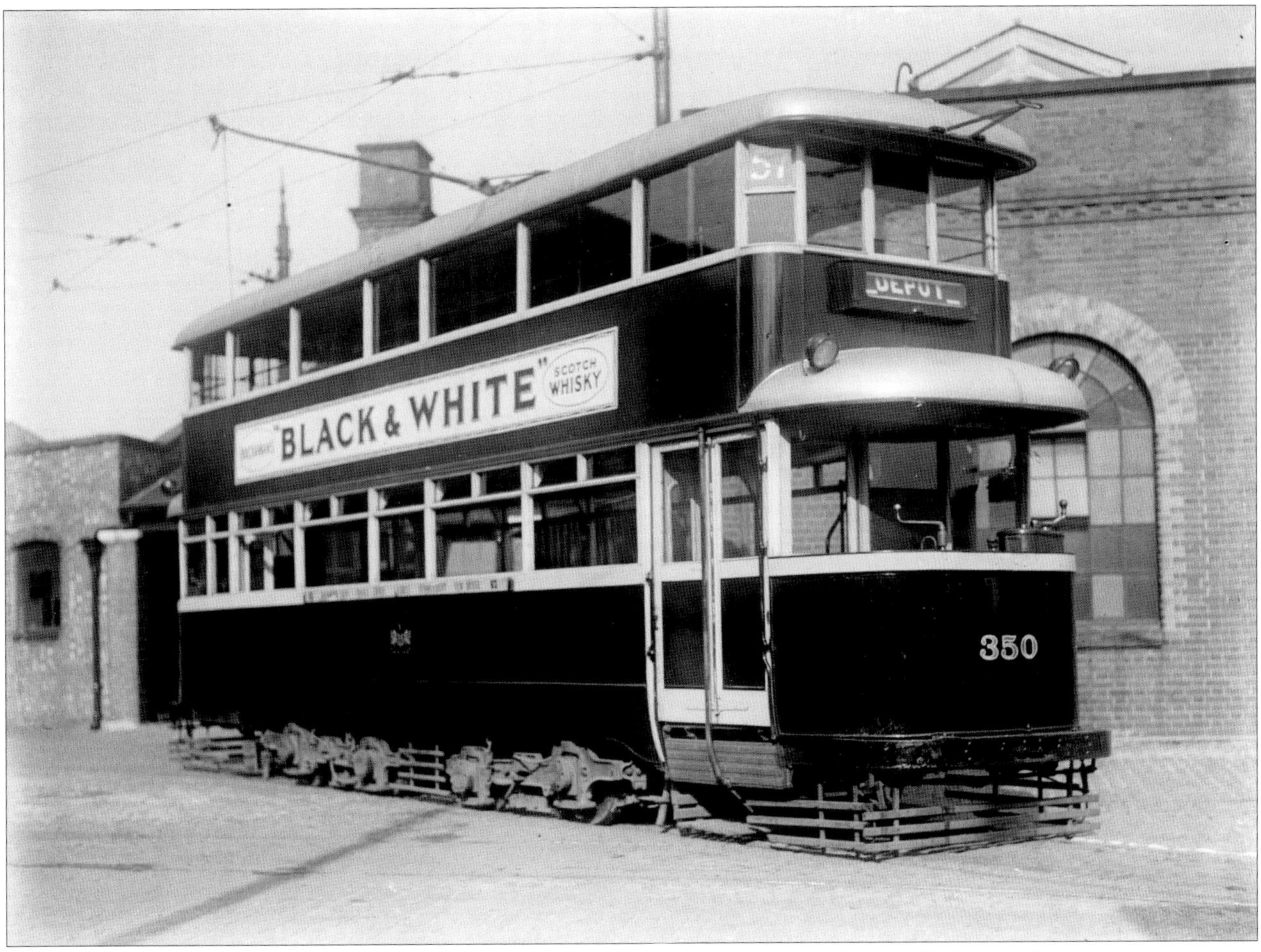

London United Tramways experimental tramcar 350, also known as 'Poppy', date unknown. (H Nicol/National Tramway Museum)

Metropolitan Electric Tramways 320, date unknown. (H Nicol/National Tramway Museum)

Turning to London County Council's tramway, by the late 1920s it had also begun an ambitious refurbishment programme that involved the upgrading of many of its existing E/1 tramcars, a process that was referred to as 'Pullmanisation' [21]. In addition, 150 much more powerful metal-bodied tramcars - the E/3s and HR/2s – were introduced in 1929 and 1930 [22].

In addition to this modernisation programme, the tramway had also embarked, in September 1929, on a major infrastructure project that involved deepening the Kingsway Subway, which formed the only link between its northern and southern tramway networks, with a view to accommodating double deck tramcars and thus increasing the capacity of its cross-river services [23].

This was the backdrop against which LCC No. 1 was commissioned and launched; and yet, as has been seen, unlike the examples of successful tramway rejuvenation projects in Britain and abroad, Bluebird neither facilitated the introduction of a successful new class of tramcar nor helped stave off the programme of tramway conversion and abandonment that commenced just three years after its highly acclaimed launch.

In accounting for its failure to fulfil the hopes that had evidently been vested in it, two inter-related issues need to be addressed. The first, and most obvious, factor has to do with location, since many of the geographical and spatial attributes that had contributed to successful tramway revivals elsewhere were not shared by London County Council's tramways.

A second factor which compounded this had to do with timing, since the period between LCC No. 1's inception and launch coincided with a number of major external financial and political events that also had a profound impact on transport policy developments in London. These related issues will be addressed in Chapter Two.

These challenges did not occur in a vacuum, however, and their impact was undoubtedly influenced by a host of other factors that had developed over the preceding decades. They include: the historical context within which London's public transport systems - including its tramways - had taken shape; the legislative and regulatory framework within which these systems operated; the organisation and responsibilities of local government and the impact this had on local transport policies; the competitive strategies of rival public transport operators across London; and, finally, the strategic decisions taken by the LCC and its tramway department in response to the opportunities and threats they were faced with. These underlying contextual issues will be investigated more fully in Chapter Three.

Chapter 2

Revival hopes derailed: challenging location and unfortunate timing

Jim Dignan

Challenging location: the public transport context in London

Other cities may have experienced a tramway renaissance during the inter-war years, but it by no means follows that the adoption of similar strategies, based on the introduction of modern tramcars and investing in infrastructure, would necessarily work elsewhere. London, in particular, faced a number of unique challenges in seeking to emulate their success. During the first quarter of the twentieth century, for example, London not only had the largest population of any city in the world [1], but also the widest urban sprawl (see below). London was already the largest city in the world by 1800, and over the course of the next century its population grew rapidly, reaching over seven million by 1914. It was only displaced, by New York, in 1925.

In addition to its geographical spread and population size and density, the city was also bisected into two roughly equal halves by a sizeable river with a limited number of crossings and incessantly voluminous traffic densities [2]. Both of these factors posed a formidable challenge to public transport operators in the capital.

Moreover, these physical challenges were compounded, in London's case, by a series of equally problematic administrative ones. London County Council was created by statute shortly before the dawning of the electric tramway era [3]. The boundaries of the newly created County of London, which it administered, encompassed most of the London conurbation at the time, which posed an immediate challenge since the area they encompassed was greater – in relation to population density – than that of any other city in the world during that era [4].

However, the powers the LCC exercised, including those relating to highways and public transport, were not unfettered. This posed another considerable challenge as a second tier of local government, comprising 28 Metropolitan Boroughs (established by the London Government Act, 1899), acted as the designated 'highway authorities' within their own respective borough boundaries.

An additional administrative body, separate from, and also pre-dating, both the LCC and the Metropolitan Boroughs, was (and still is) the City of London which encompasses the historic centre of the city and also its hugely influential central business district.

Although encompassing a tiny portion of the London metropolis – just over a square mile, while the LCC administrated 117 square miles and Outer London had an upper limit of 605 square miles [5] - the City of London is nevertheless both a discrete ceremonial county (the smallest in the United Kingdom) and also a separate local authority that again exercises a unique set of powers in its own right.

These complex administrative arrangements had important ramifications for the state (and ultimately, perhaps, even for the fate) of London's tramway systems. One obvious consequence is that, in sharp contrast to most of the examples of successful tramway rejuvenation programmes of the era, there was at this point no single body with overarching responsibility and authority for developing a coherent and comprehensive public transport network for the capital. What had evolved instead was a rather haphazard patchwork of largely fragmented and frequently disjointed local tramway systems that were operated autonomously under various forms of ownership and control.

By far the largest of these was LCC Tramways, which by 1932 operated 1,713 tramcars over 167 route miles, constituting around half the total route mileage and two-thirds of the total number of tramcars that were absorbed by the LPTB the following year. This system had initially been established through the acquisition of three distinct local horse tramway networks that were purchased from their former private operators. Two of these, the North Metropolitan Tramways Company and the London Street Tramways Company, were located to the north of the river Thames while a third system, the London Tramways Company, was situated on the south bank of the river. There were no physical connections between them to start with and it was only possible to establish connections between the first two networks following a merger in 1897.

The map of the LCC Tramway in 1928 highlights two of the network's principal operational challenges, the first being the virtual separation of the 'North Side' and 'South Side' networks. Prior to the opening of the rebuilt Kingsway Subway in 1931, double deck trams from the northern sections were only able to access the LCC's central repair and maintenance depot at Charlton in South East London by arrangement with the company-owned Metropolitan Electric Tramways (MET). Even more debilitating for the tramway was the large doughnut-shaped hole at the centre of the network which resulted from the persistent refusal of the boroughs of Kensington and Chelsea, Westminster and also the City of London to countenance the running of tramcars through their streets (an issue that will be explored further in Chapter 3).

This effectively established a 'no go' area for tramcars across the central business district and commercial hub of the city, depriving operators of the potential benefits of 'through-running' services across the network, not to mention the revenues that might have been generated from these prosperous districts and their extensive retail and employment outlets. It also afforded a 'free pass' to the privately owned bus companies who were at liberty to ply for passengers within the zone without fear of competition from municipally owned tramcars.

In addition to the extensive LCC Tramways, in 1932 there were also a further ten municipally owned and operated systems, nine of which lay to the east and one to the south of the LCC network. By the end of 1928 they included Barking (which abandoned its tramway in favour of motor buses in 1929), Bexley, Dartford, East Ham, Erith, Ilford, Leyton, Walthamstow and West Ham to the east and Croydon to the south [6]. London was also home to three privately owned and operated systems, often referred to as 'company tramways', comprising Metropolitan Electric Tramways (MET) in the north and north west of the city, London United Tramways (LUT) in the west and south west, and the South Metropolitan Electric Tramways & Lighting Co. (SMET) in the south of the city. A further complication lay in the fact that both Middlesex County Council and Hertfordshire County Council also owned track which was leased to MET.

Unlike the tramways, most but not all of which were municipally owned and operated, London's bus services had traditionally been the exclusive preserve of private operators, and Parliament refused

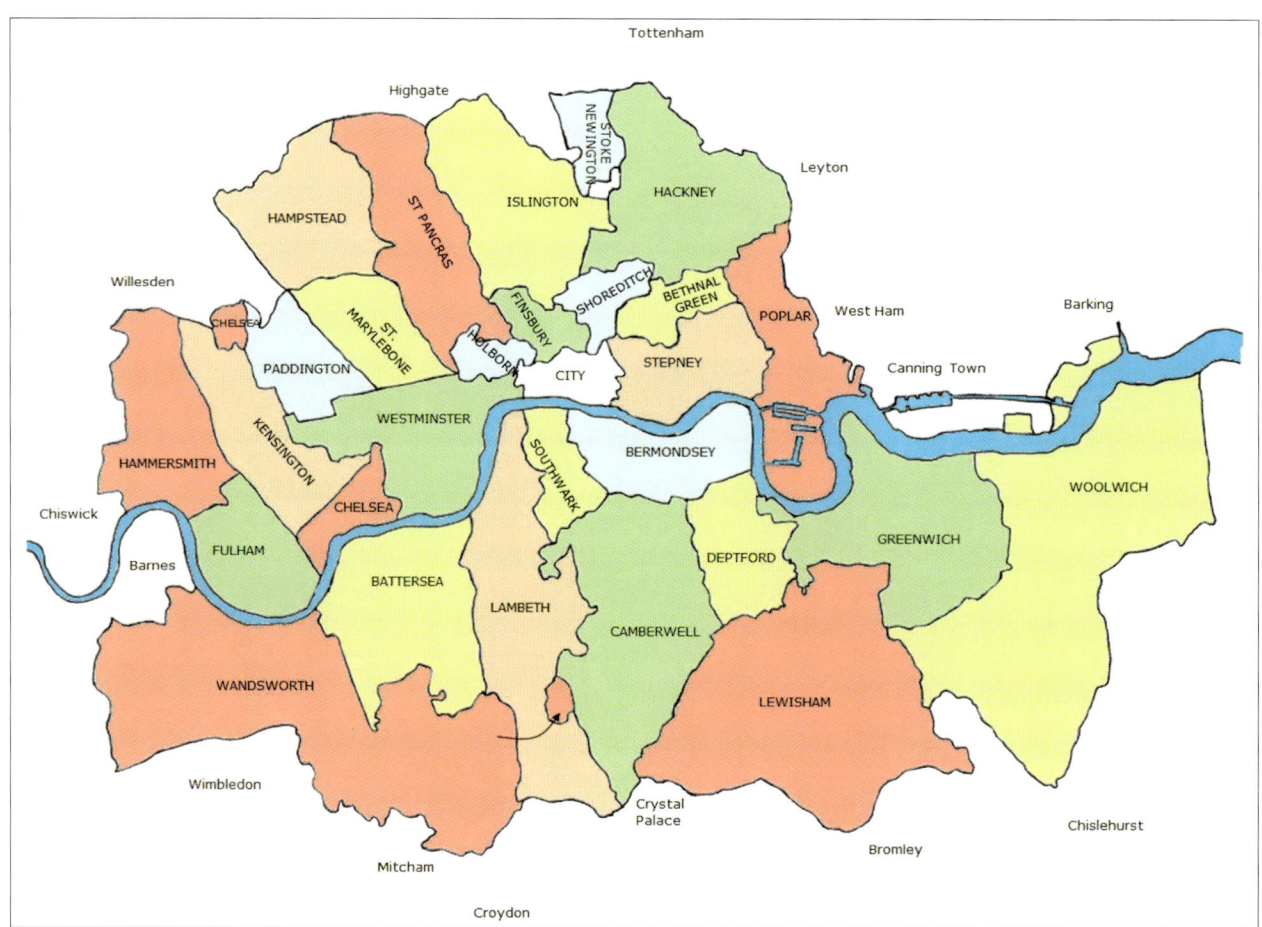

The Metropolitan Boroughs that comprised the County of London. (www.socialhousinghistory.uk)

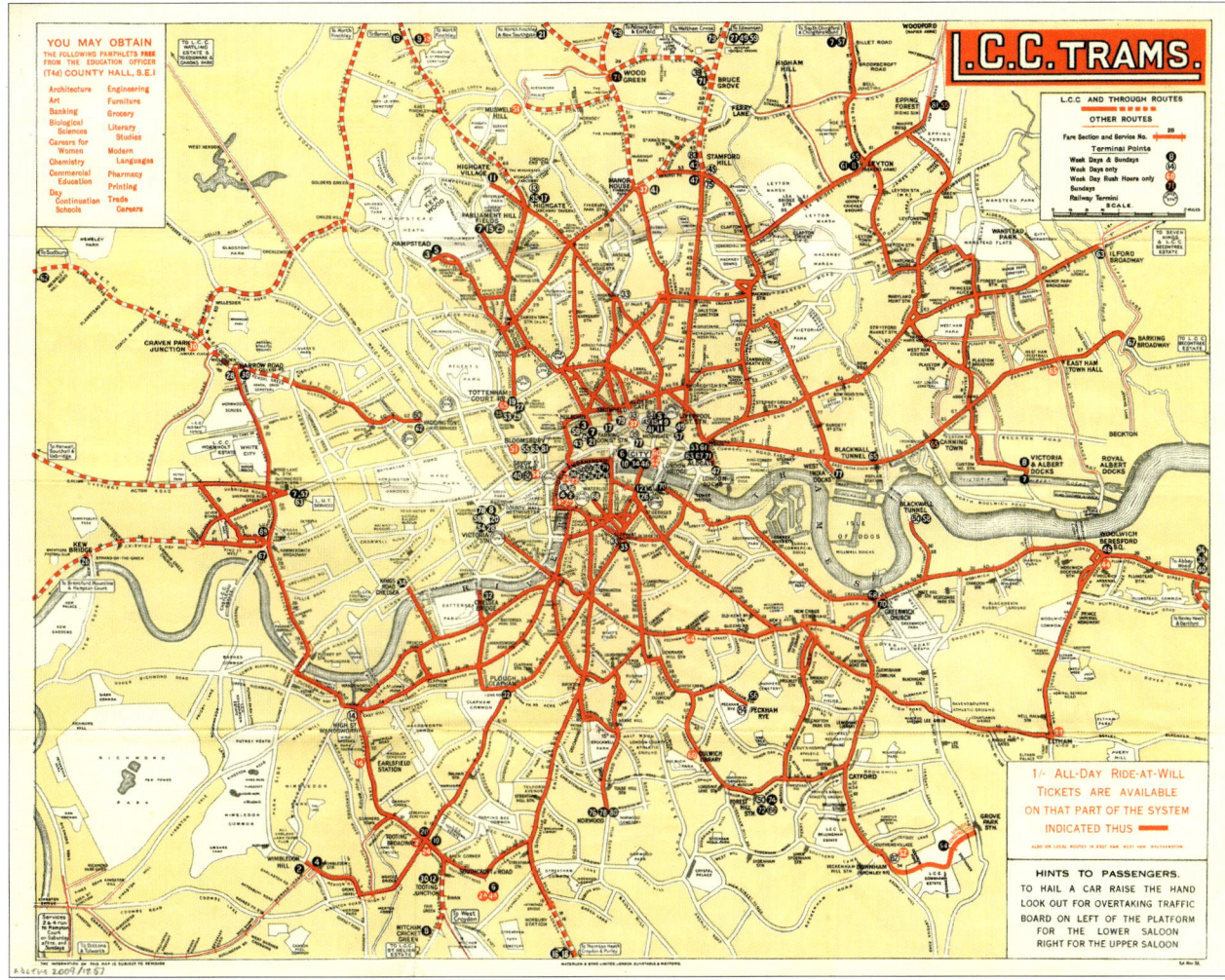

Pocket tramway map and timetable, issued by London County Council Tramways, May 1928. (© TfL from the London Transport Museum Collection)

to authorise municipal tramways in London to run their own bus services [7] even though this was commonplace outside the capital. One final twist in this complex saga is that the municipal tramways found themselves at an increasingly competitive disadvantage as, over a period of time, a powerful alliance of privately owned public transport undertakings brought together underground railway operators, bus companies and also the three company tramway undertakings under the control of a single entity - the Underground Electric Railways Company of London Limited (UERL) - which is sometimes referred to as 'The Combine'.

This was the inherently unstable state of the tramway system in London around the time of Bluebird's inception. However, worse was to come in the wake of an unparalleled series of financial and political crises that came to a head at this precise juncture and would ultimately usher in a new era for London's entire public transport system.

Unfortunate timing for LCC No. 1's inception: the makings of a 'perfect storm'

The first specific reference to the LCC's proposal to develop a completely new type of tramcar was attributed to the Highways Committee on 24 October 1929, when expenditure up to a maximum of £5,000 was allocated for this purpose [8].

With singularly unfortunate timing, this was also the day of the Great New York Stock Market Crash which, following a similar crash on the London Stock Exchange in the previous month, was to usher in a 12-year period of economic and financial turbulence around the world. This period has come to be known as the Great Depression and was marked by rising unemployment and rapidly falling living standards across much of the social spectrum.

By the summer of 1931, the British economy was entering a critical phase. An impending budget deficit of £120 million was compounded by a dramatic run on the pound on 11 August; foreign investors scrambled to remove their money from the City of London in response to concerns that ill-advised speculation had left bankers exposed to foreign liabilities of over £400 million [9].

These rapidly intensifying economic crises in turn precipitated a parallel political crisis as the minority Labour administration that had been in power, with Liberal Party support, since May 1929 was unable to agree on an effective response to the growing financial turmoil.

This prompted Ramsay MacDonald, the Labour Prime Minister, to tender his resignation in August 1931, resulting in a government of national unity, which chiefly comprised the Conservative and Liberal Parties with Ramsay MacDonald still at the helm; the bulk of the Labour Party and some Liberals went into opposition at this point. A further General Election in October 1931 resulted in a landslide win for MacDonald's National Government - still dominated by the Conservatives - which remained in office until 1935.

These interlocking economic and political crises were serious enough in themselves to have had a severely detrimental impact on the prospects for LCC No. 1 and any plans the LCC might have harboured for constructing a new class of Bluebird-inspired tramcars to emulate the Feltham trams that were operated by its company-owned rivals, MET and LUT.

For many years, the LCC had sought to hold down tram fares in pursuit of a long-standing housing policy designed to encourage residents to relocate from congested inner-city areas to suburban districts. In 1921, such policies had contributed to a deficit crisis for the LCC that had resulted in an application for a fare increase. A decade later, any major investment in fleet renewal would almost inevitably have required significant borrowing on the part of the LCC [10].

One of the first effects of the Great Crash was the imposition of severe curbs on public expenditure and also (at least in the short term) high interest rates, which would have made such borrowing much more expensive. Interest rates rose from 4.5% to 6% around the time of the Great Crash, in 1929, before falling to 3% in the second half of 1930; but they again doubled to 6% during the winter of 1931-2 [11]. Moreover, all tramway operator loans had to be sanctioned by the Board of Trade (subsequently the Ministry of Transport), which meant that central government could influence local authority transport policies especially during periods of economic turbulence [12]. An additional restriction to which the LCC was subject required it to submit an annual Money Bill to Parliament, which meant that MPs' approval routinely had to be obtained for additional items of expenditure, including loans [13].

Although the company-operated Feltham trams were being constructed at around the same time as

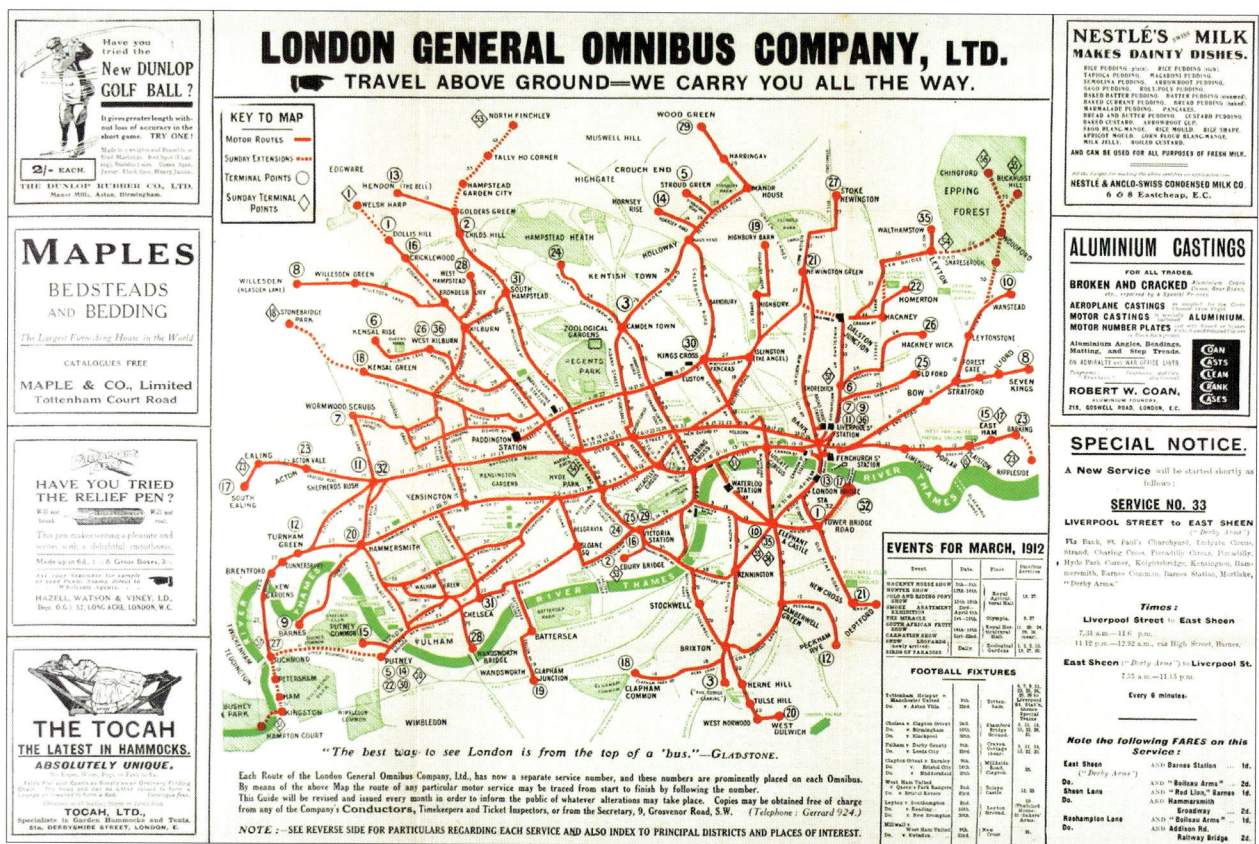

A map of the routes served by the London General Omnibus Company in 1912 shows a much more even coverage of the central London area than the corresponding LCC Tramways map above.
(© TfL from the London Transport Museum Collection)

LCC No. 1, they were fortunate in escaping any direct adverse consequences flowing from the Great Crash. The cash-strapped LUT, for example, had authorised the acquisition of its batch of Feltham trams on 13 February 1929, just over six months prior to the Crash [14], while MET financed its Felthams from the recent sale of its shares in the Northmet power company [15].

Quite apart from these economic constraints, which would clearly have impinged on the LCC's ability to formulate and pursue its own policy initiatives and priorities, the resulting political upheaval was to transform the overall public transport policy context in an even more radical manner. Prior to these tumultuous events, it is fair to say that London's public transport system had developed over the years in a piecemeal, uncoordinated fashion and, in the absence of any city-wide transport authority for the capital, with little or no strategic planning or oversight. As a result, various modes of public transport – trams, buses, underground and main line railways – had been developed by a wide variety of operators, including privately run companies and municipal undertakings. This was largely an era of unbridled competition and even 'empire-building' as various transport undertakings sought to gain and retain a dominant market share at the expense of their rivals.

Some of these, notably the London General Omnibus Company (LGOC), had a city-wide remit and enjoyed a virtual monopoly in the provision of bus services throughout London. Others, including no fewer than 12 tramway undertakings, provided tramcar services which for the most part operated exclusively within particular geographical areas. Most of these were owned and operated by individual municipal boroughs.

As we have seen, however, three of these undertakings were company-owned, and two of them - MET and LUT – had collaborated in the development and production of the Feltham tramcar. An additional complication was that, at that time, the three company-owned tramways formed part of the aforementioned wider transport grouping - the Underground Group, or UERL - which not only owned most of the London Underground network but also, as will be described more fully below, the LGOC, which it had acquired in 1912.

It is clear from this brief analysis that, unlike bus and underground railway interests, the tramway sector did not speak with a single unified voice; nor could the various tramway undertakings be expected to develop a shared perspective or agenda. Moreover, the Underground Group had by this stage assumed a dominant position with regard to all other transport operators including the municipal tramway sector, notwithstanding the relative size and influence of the LCC.

By the mid-1920s, however, various public transport policy changes were already under way. For example, the first, short-lived Labour government's London Transport Act of 1924 had restrained bus competition by limiting the number of buses permitted to operate on certain designated streets. As most major tramway routes were designated in this way, London's tramways were offered a degree of protection on this front, at least for a time.

London's tramways were not the only beneficiaries of this legislation. In fact, it was adopted in response to a strike by London's omnibus crews, which resulted in their employer acceding to their demands

Early logo for the UERL produced in 1907 before the acquisition of the London General Omnibus Company. (© TfL from the London Transport Museum Collection)

1933 LPTB poster showing the transfer of responsibility for various transport undertakings to the London Passenger Transport Board. The poster also features the LPTB's new logo, designed by Cecil Bacon in the same year. (© TfL from the London Transport Museum Collection)

in return for an undertaking that action would be taken to curtail harmful competition by small-scale 'pirate' bus operators. The Act also established the London and Home Counties Traffic Advisory Committee which, in 1927, recommended the better co-ordination of public transport in London, but with ownership remaining with existing proprietors. Indeed by 1929, Bills promoted by the LCC and London Electric Railway Company that were intended to give effect to these proposals had been passed by the House of Commons and were due to go to the House of Lords. Had they been enacted, the LCC would have continued to operate as a provider of public transport and to remain in charge of its own destiny with regard to the running of its tramway [16].

This was the context in which the minority Labour government came to power in 1929. In terms of public transport policy, it flatly rejected the hitherto dominant idea of unrestrained competition. However, the incoming transport minister, Herbert Morrison, also favoured the adoption of a much more coordinated and radical approach to public transport policy-making for the capital.

In his view, the best way of achieving this was by means of a single transport undertaking to manage the entire public transport sector. He favoured the establishment of a publicly owned board, operating the public transport network independently as a commercial enterprise, guided by the principles of efficiency and profitability [17].

On 23 March 1931, a Bill was introduced to give effect to Morrison's proposals by establishing a London Passenger Transport Board (LPTB) to take control of London's various passenger transport undertakings. This Bill survived the demise of Ramsay MacDonald's Labour government in October 1931, having been adopted by the incoming National Government, and became law on 1 July 1933, just over a year after LCC No. 1 had entered public service on 10 June 1932.

This event was of crucial significance for the prospects, not only of LCC No. 1 itself, but also those of London's tramways in general, as responsibility for determining overall transport policy passed from the various transport undertakings, including the LCC, to the newly constituted LPTB. Whereas Morrison had initially envisaged that the Board would operate under municipal leadership, he gave way on this in order to win political support for the plan, and agreed instead that a majority of directors should consist of representatives drawn from the private sector [18].

Crucially, perhaps, the first chairman and vice-chairman of the first Board were Lord Ashfield and Frank Pick, both of whom had held similar positions with the Underground Group that had been the LCC's principal rival for many years. Of the Board's other five members, only two had previously served on the LCC but neither had had any direct involvement with the tramways undertaking [19].

Meanwhile, shortly after the London Passenger Transport Bill had been published, another event occurred which heralded a further significant change in the direction of public transport policy in the United Kingdom. A Royal Commission had been established in 1928 to investigate the problems arising from the growth of road traffic and to consider what measures should be adopted for its more effective control. In April 1931 it published its Report.

After assessing the strengths and weaknesses of trams as a mode of public transport, the Commission uncritically accepted the conventional wisdom of the day that tramcars were a major cause of urban transport congestion, concluding that Britain's tramways were in a state of obsolescence. This assessment underpinned its recommendation that no new tramways should be constructed and that existing tramways should gradually be phased out and replaced by other forms of transport [20].

Counter arguments setting out the case for tramway retention and calling for long-standing grievances to be remedied were advanced by representatives of the LCC, the Municipal Tramways & Transport Association, the Tramways & Light Railways Association and the Transport and General Workers' Union. These, however, were summarily dismissed by the Commission, most of whose members (at least seven out of twelve) were, or had been, Conservative or Unionist politicians compared with two Labour MPs who formed part of the small minority of members (three out of twelve), without titles.

Although the Report didn't have an immediate impact on either national or local transport policy, it probably reflected the consensus shared within transport policy circles and the public alike. So, it is not altogether surprising that in 1934 the LPTB announced a plan to phase out its tramcars; and in October 1935 it launched a programme to progressively replace them with trolleybuses [21].

To summarise, since work had commenced on the construction of LCC No. 1 just over five years previously, a combination of circumstances, including a period of severe financial and political turbulence, had brought about a dramatic shift in the organisational framework for operating and developing London's public transport sector, together with an equally dramatic change in the direction of public transport policy with specific regard to London's tramways.

Faced with an upheaval on this scale, it was probably inevitable that LCC No. 1 would remain a one-off, however successful its design and regardless of its potential capacity to counter the competitive challenge posed by buses. In order to understand how these external financial and political events could have had such a far-reaching impact on London's transport policy, however, it is necessary to delve a little deeper into the events and circumstances which placed the LCC's and London's tramways in general in such a vulnerable position.

The dramatic shift in transport policy owes at least as much to the effect of several long-standing structural handicaps to which tramways were uniquely susceptible, as to the tumultuous events of the era surrounding LCC No. 1's inception. The cumulative effect of these handicaps was to acutely and progressively disadvantage London's tramways specifically by preventing them from being able to compete on equal terms with rival modes of transport.

The handicaps were not inevitable but arose from a series of legislative, administrative, political, operational and fiscal constraints which greatly impeded the development of tramway operations in London over many years. Various attempts were periodically made to address these constraints. Had they succeeded, the direction of London's transport policy might well have taken a different turn; but, as will be seen, they were not successful. As a result, LCC No. 1 was destined to remain a unique pointer to an alternative transport policy direction that might otherwise have been followed.

Frank Pick (left) and Lord Ashfield photographed at Acton Depot circa 1923. (© TfL from the London Transport Museum Collection)

Chapter 3
Revival hopes frustrated: underlying obstacles
Jim Dignan

Operational and regulatory constraints

All forms of public transport are subject to operational constraints of one kind or another and these can have a major impact on their commercial viability and ability to compete with rival transport modes. Even the simplest of horse-drawn tramways, for instance, depended on the installation of complex, disruptive and expensive infrastructure in the form of track-work, stables and depots before operations could commence. Rival omnibus operators did not face such serious problems.

Even greater challenges accompanied the subsequent adoption of electric traction, which required the additional installation of generating equipment and either overhead wiring or alternative mechanisms for the transmission of electric current to the tramcars. The former option raised aesthetic considerations while the latter presented further technical and also additional financial challenges.

Tramways were not unique in this respect. The adoption of surface and underground railways also entailed complex and expensive infrastructural challenges. Unlike these other modes of public transport, however, most urban tramways were designed to operate for much of the time on public thoroughfares and were thus obliged to share their running tracks with other users including pedestrians, horses and, subsequently, other forms of vehicular traffic.

The challenges were not merely technical – the development of grooved rail that was flush with the surface of the street, for example – but also regulatory, since some form of accommodation had to be reached between the various interests at stake. These included: the tramway operators themselves; the travelling public they were intended to serve; the relevant highway authorities; those with business premises adjoining a proposed route; and any rival transport operators, including those operating in the private sector where shareholders' profit-seeking motives formed yet another interest group.

As a result, tramway operators throughout the country were entirely dependent on the regulatory framework within which they were required to operate, as enacted by Parliament. The key legislative development was the Tramways Act of 1870, which established the principle that local authorities had the power to authorise the operation of tramways within their own geographical areas. However, this power was tempered by several restrictive financial and other impediments that were also set out in the Act, the effect of which was to place tramway operators at a distinct competitive disadvantage compared with those operating rival modes of transport.

Among the many financial constraints to which tramway operators were uniquely subject was a provision in the Act that rendered tramway operators solely liable to pay for the construction and ongoing maintenance of the permanent way plus 18 inches on either side of the tracks [1]. In addition, municipal rates were levied on the entire surface area of the road for which tramway operators were responsible, effectively imposing an additional tax on tramway operators to which rival bus operators were not subject.

In practice, this usually meant that, in the case of double track systems, tramway operators were obliged to pave the entire road even though the thoroughfare was used by other vehicles, including those run by rival public transport operators, notably bus companies. Moreover, this was a burden that became more onerous as road traffic increased in density and vehicles became larger and heavier. London tramway operators were particularly disadvantaged in this respect as the density of the built-up area in the capital made it much more difficult to construct dedicated reserved tracks of the kind adopted in many provincial cities, which were not shared with other modes of transport (see Chapter 1 for examples).

In addition, tramway operators were obliged - before a Provisional Order was granted under the terms of the 1870 Act - to satisfy the Board of Trade that an adequate, regular and comprehensive service would be provided at fare rates that the latter deemed necessary and proper [2], and were further required by statute to offer heavily subsidised workmen's fares. No such obligations were imposed on rival bus operators, which were consequently at liberty to 'cherry pick' the most lucrative routes and also the most favourable times at which to operate services, with a view to maximising their financial returns while minimising their outlay.

The first schemes to be sanctioned under the 1870 Act were horse-drawn tramways promoted by private companies and funded by the sale of shares to investors. Local councils were not initially authorised to operate such tramways themselves, although they were given an option to purchase and take over these operations after a period of 21 years [3].

Delegating the power to authorise tramway operation to local authorities in this way rendered the future development of tramways within a given area hostage to prevailing local political sentiments. This might not be particularly problematic within a self-contained local authority but could, and did, cause serious problems in a conurbation the size of London, comprising, as we have seen, large numbers of relatively small geographical units. These did not always see eye to eye with one another, and decisions on tramway development within a given area might be influenced by purely local considerations and vested interests, without any responsibility for promoting or meeting the transport needs of adjacent areas or, indeed, of London as a whole.

This was less of a problem in the days of horse-drawn tramways, which tended to be relatively restricted in length, but it became much more of an issue when the advent of electric traction made it possible to greatly enlarge tramway systems and develop much longer routes. Such expansion could only occur, however, if agreement could be secured from the relevant local authorities through which the lines were to run.

In the event of a disagreement between the relevant highway authorities, a tramway could only proceed if at least two-thirds of them were in favour of a proposed development [4]. The sheer number and diversity of local authorities involved did indeed cause acute problems in London's case, as will be seen in the following section.

An additional constraint was a provision in the 1870 Act extending a right of objection to the owners or occupiers of any premises, whether domestic or commercial, where the clearance between the pavement edge outside their premises and the nearest tramway rail was less than 9 feet 6 inches [5]. Given the relatively narrow thoroughfares in most urban areas, such objections occurred frequently, and were conclusive if one-third of the owners or occupiers objected to the tramway being laid.

Although a late arrival on the London tramway scene, having been established in 1889 as a directly elected municipal authority whose remit covered most of the inner London area, the LCC was nevertheless subject to the same regulatory framework. However, the LCC's powers did not displace those enjoyed by the 28 constituent metropolitan London Boroughs (let alone those of other tramway operators such as Walthamstow, a municipal borough in the County of Essex) which continued to discharge many functions (including acting as highway authorities) within their own jurisdictions.

Consequently, having developed a fragmentary tramway network through the acquisition of existing privately owned horse tramways, the LCC was unable to develop new tramways, or even extend its existing ones in order to link them up, without the approval of the relevant borough councils, since it was not a highway authority in its own right.

Other tramway services, as we have seen, were operated by company-owned systems based outside the LCC County boundaries. As there was usually no physical connection between the in-county and out-of-county services, no through running was possible; passengers had to switch cars at such locations, which were often situated in the middle of busy urban streets. This resulted in delays, inconvenience and additional expense - due to the absence of through-ticketing arrangements - for passengers, and greatly increased congestion for other road users. Even as late as 1923, the lack of physical connections between different parts of the wider London tramway network still had not been satisfactorily resolved in all districts.

An even bigger problem was the LCC's inability to extend its tramway operations into parts of London which either steadfastly resisted any incursions down the years, or else imposed expensive stipulations on the way the tramway was to be constructed and operated.

This restrictive regulatory framework was to prove a major stumbling block which the LCC was unable to fully circumvent despite many attempts over the years. As a result, it was never able to develop the kind of efficient and comprehensive tram-based public transport network which might have been expected in a city the size of London, and which could perhaps have enabled the capital's tramways to better withstand the increasing competition from rival modes of transport.

Municipal parochialism and obstructionism

Municipal involvement in tramway operations had both benefits and drawbacks. One of the benefits was that local authorities were able to rapidly finance the conversion of small-scale company-owned horse tramways to electric traction and to greatly expand the size and density of their tramway networks if they so wished. However, they also possessed an absolute power to veto any developments they opposed, without right of appeal, which enabled them to prioritise their own interests without regard to any impact on the wider travelling public.

Historically, the boroughs of Kensington, Chelsea, Westminster and the City of London had long opposed the development of horse trams in their areas [6]. Indeed, their implacable opposition dated back to the dawn of the tramway era prior, even, to the implementation of the 1870 Tramways Act, which was the source of their veto powers. Their antipathy may well have been kindled by the controversial endeavours of one of the earliest pioneers and promoters of horse-drawn tramways, George Francis Train, who was a flamboyant and eccentric American entrepreneur.

An early attempt to introduce a horse-drawn street tramway in central London, by the London General Omnibus Company in 1857, had foundered in the face of strong resistance from municipal interests and also from Sir Benjamin Hall, MP for Marylebone and Chief Commissioner of Public Works. The Bill - to construct lines from Notting Hill Gate through Bayswater to the Bank, via New Road, City Road and Moorgate with a branch from King's Cross to Fleet Street via Farringdon Street – had been defeated at its second reading [7].

Despite this, Train, having successfully inaugurated Britain's first street tramway in Birkenhead in 1860, turned his attention to London, securing approval for the construction of three horse tramways in the heart of the capital, though all three ventures proved to be short-lived [8]. The first line, which ran down Bayswater Road, was opened on 23 March 1861 but closed after six months. The second line, which ran down Victoria Street, opened on 15 May 1861 but was taken up on 7 March 1862. A third line, which ran from the south side of Westminster Bridge to Kennington Gate, opened on 15 August 1861 but was removed on 21 June 1862 [9].

George Francis Train, photographed around the time he was building his tramways in Britain. (www.npg.si.edu)

Although they were popular with passengers, Train's pioneering tramways suffered from a serious design fault stemming from the use of a step rail system that projected above the surface of the surrounding roadway, to the unsurprising detriment and inconvenience of other road users.

Shortly after the opening of the third line, a fatal accident occurred in which a boy died after being hit by a tram [10]. Train was arrested and spent some time in prison before being acquitted when the case came to trial. However, he was then charged at Kingston Assizes with the offence of 'breaking and injuring' Uxbridge Road. The court determined that in the absence of an Act of Parliament, he was not authorised to lay the tracks and fined him £500, ordering him to dismantle the tramway. When he refused to pay the fine or dismantle the tracks he was put in debtor's prison. Although he subsequently tried to secure Parliamentary approval for his lines, he was unsuccessful and so the tracks were removed.

Train's horse tramways were popular with the travelling public. However, the lasting antagonism engendered by his somewhat impetuous and reckless attempts to demonstrate the tramways' potential in what was then the largest city in the world might ironically have blighted the longer-term prospects for tramway developments in the capital [11].

An artist's impression of what Train's first London tramway of 1861 would have looked like. It is in Bayswater Road showing Marble Arch. (National Tramway Museum)

Horse tram 'The People' at the eastern end of Victoria Street, on the single line section opened by George Francis Train in 1861. This second London tramway, like its predecessor in Bayswater Road, did not last long because of the rails protruding above the surface of the road.
(© TfL from the London Transport Museum Collection)

Opposition continued into the era of electrification and, as we have seen, resulted in the virtual exclusion of tramway operations from most of the central business district and also the West End – potentially the most lucrative districts of central London. The LCC, therefore, had to contend not just with the barrier presented by the River Thames, but also the existence of a sizeable doughnut-shaped void in the centre of their network, together with the operational constraints, expense and inconvenience that must have resulted from the absence of effective connections between different parts of the LCC's extensive network. This not only prevented the Council from penetrating the inner core of the city but also precluded the development of many cross-city services, as featured in numerous successful large metropolitan tramway systems around the world.

The only exception was a single route through the city, made possible by the construction of the Kingsway Subway in 1906-8. Initially this could only be used by single deck cars but, following an enlargement, it was eventually capable of accommodating double deck tramcars after it was reopened in January 1931. This remained the sole connection between the LCC's North and South London tramway operations which, in other respects, formed two virtually discrete networks for the entire duration of tramway operations.

The Royal Commission on London Traffic in 1905 reported that tramway development in London had undoubtedly been seriously checked by the operation of these municipal vetoes, which were sometimes exercised without due regard to the interests of the general public and the importance of establishing through tramway communication [12].

Other boroughs were more favourably disposed to tramcars operating on their streets but, even among these, a number of inner London boroughs, including Croydon, Hackney, Lambeth, Lewisham, Stepney and Wandsworth, objected on aesthetic grounds to the erection of overhead wires [13]. Their power of veto obliged the LCC to install an alternative and much more expensive form of current collection, using an underground conduit system between the two running rails, which fed power to the tramcars by means of a 'plough' suspended from each car.

Consequently, at various points on the system tramcars had to switch their method of current collection from conduit to overhead, or vice versa. This necessitated the inclusion of a 'change pit' by the side of the track, which is where the tramcar relinquished, or picked up, its plough. By the time the LCC tramway was fully developed more than 20 of these change pits were in use [14].

Opened in 1906, Kingsway Tram Subway once ran underneath the length of Kingsway, taking passengers from Holborn to Aldwych and emerging below Waterloo Bridge. (National Tramway Museum)

Conduit systems were not only nearly three times as expensive to install as overhead wiring, but also significantly increased the cost of

Attaching a plough to a tramcar at Wandsworth change pit, 1933. (H Nicol/National Tramway Museum)

LCC No. 1 at Manor House change pit, date unknown. (A W Bates/National Tramway Museum)

both track and tramcar maintenance. Ploughs were very short-lived and required frequent replacement [15]. Although some boroughs did eventually agree to overhead wires - routes thus equipped included those serving Amhurst Park, Downham, Eltham, West Norwood, Woolwich and Victoria Park [16] - the need for tramcars to operate on different parts of the network meant that many had to be dual-fitted with equipment adapted for both forms of current collection.

The combined effects of such operational constraints and an imperfect regulatory regime led to Londoners being much less well served by trams than passengers in other parts of the country, where the tramway networks were three or four times as dense as they were in the capital. One way of measuring track density is with reference to the number of inhabitants per route mile of operational tramway in a given area. Figures for the end of 1904 revealed that whereas many provincial cities had between 10,000 and 15,000 inhabitants per operational route mile, in London the corresponding figure was 33,661. The 'densest' tramway network at the time was in Manchester with 8,937 inhabitants for every operational route mile, though even this figure was exceeded by a number of cities in the United States and Continental Europe [17].

Failure to physically 'join up' the various disconnected or autonomous tramway systems in the capital also had a detrimental impact on the potential capacity and efficiency of the tramway. The delays caused by large numbers of 'dead-end terminals' at municipal or county boundaries were estimated to have reduced by half the number of tramcars per hour - from 150 cars to 72, according to estimates by the Advisory Board of Engineers [18] - capable of operating in any one direction.

The constraints which handicapped the development of the tramway network to its full potential were widely recognised at the time, and various attempts were made to address them over the years.

One of the most obvious solutions would have been to establish a co-ordinating authority with responsibility for ensuring the rational and comprehensive development of London's public transport infrastructure. Many attempts were made to achieve this objective in the decades preceding the creation of the LPTB in 1933 and, had they succeeded, London's public transport infrastructure might well have developed along very different lines. All such attempts foundered, however, in the face of Parliamentary prevarication or outright intransigence.

Failure to reform the regulatory framework

As long ago as 1855, with traffic congestion already becoming a serious problem, a Select Committee of the House of Commons was set up to consider communications within the metropolis. Much of the congestion resulted from a plethora of horse-drawn cabs, carts and other vehicles including horse-drawn omnibuses which, at that time, were the only means of conveying the growing number of passengers brought into the centre of London by main line railways or river steamers [19]. It advised that Parliament would not be able to remedy matters in the absence of an authority with the powers to plan and carry out improvements to meet the capital's existing and prospective travel needs [20]. This recommendation had not been acted upon by 1861, when the first horse tramways began appearing on the streets of London.

The original procedure for establishing new tramways was very cumbersome and costly, and involved the passage of a private Act of Parliament. An early attempt to tackle the problem of municipal obstructionism by streamlining the regulatory process was made in 1870, many years before the LCC itself came on the scene [21]. The Board of Trade introduced a Tramways Bill which, in its original form, would have greatly facilitated the development of tramways in London and elsewhere. If successful, it would have removed the need for promoters to initiate a separate private bill for each new tramway and would also have restricted the power of local authorities to veto such developments.

The timing of the measure was unfortunate, however, as the President of the Board of Trade, John Bright - an influential orator and an effective campaigner - had fallen seriously ill, forcing his retirement in December 1870. This left the key position of President of the Board of Trade vacant at a critical stage. In the meantime, the Bill was severely mauled at the Committee stage in the Commons and, as a result, the attempted procedural reform ended in failure.

Had it been successful, the LCC, and its predecessor the Metropolitan Board of Works – whose rights and obligations the LCC inherited, although these did not include the right to operate a tramway undertaking - would almost certainly have been in a far stronger position to develop a comprehensive tramway network in central London. Such a network might also have been much cheaper to operate if local authorities had lost their power to veto overhead current collection simply on grounds of aesthetics.

A typical working day in London at the end of the 19th century showing omnibuses, hansom cabs and delivery carts jostling for position. (www.thehistoryoflondon.co.uk)

Traffic congestion in Tottenham Court Road around the end of the 19th century, comprising horse buses, cabs and other horse drawn vehicles. (© TfL from the London Transport Museum Collection)

As it was, the ensuing, and greatly amended, Tramways Act of 1870 retained the requirement of obtaining Parliamentary approval before Board of Trade certificates could be awarded, and also conferred other powers of veto upon local authorities and property owners, as described above.

We have already seen that other modes of public transport, including motor buses and underground railways, were much less circumscribed in terms of route development (see also below). While the latter were likewise obliged to obtain a private Act of Parliament, they were not subject to the same veto powers wielded by local authorities or private property owners. Nor was the LCC able to intervene in the approval process. It could oppose any Bills proposed by underground railway promoters, but provided the latter were able to muster sufficient Parliamentary support, such opposition all too often fell upon deaf ears and was readily defeated. The result was an all-too-predictable proliferation of rival public transport schemes, established without any assessment of whether they were justified by a demonstrable transport need or would operate in the public interest.

In 1901, a Joint Select Committee of both Houses of Parliament, set up to examine London's underground railways, agreed that the lack of any rational planning procedure was unsatisfactory [22]. It recommended the creation of an appropriate public body capable of exercising supervision and exerting control over all proposed projects; but it expressed no preference as to whether this should take the form of the Board of Trade, a body of Commissioners or a Joint Select Committee. It also proposed that the LCC, the City Corporation and any other councils which might be concerned should be authorised to construct, or to assist in the construction of, underground railways.

Neither recommendation was adopted, however, and so another opportunity to reform the regulatory framework fell by the wayside, allowing the underground railway network in London to continue to develop on a haphazard basis [23].

A further opportunity to reform the regulatory and planning process occurred in 1905 when the Royal Commission on London Traffic was established to consider the best means of improving London's transport, both above and below ground. Also debated were the merits of establishing a central authority to adjudicate on all proposed tramway or railway construction schemes, and the powers that should be conferred on such a body.

One of the Royal Commission's main recommendations for improving London's public transport network was a major extension of its surface tramways throughout London and the suburbs [24]. The very detailed and ambitious plans for tramway expansion in central London were set out in Part II of the published report and drew on proposals drawn up by the Advisory Board of Engineers. It also advocated the linking up of the various tramway systems, the provision of through-running arrangements wherever possible between them, and the construction of cross-river tramway links across the Westminster and Blackfriars bridges. If its ambitious expansion proposals had been adopted, the effective 'no-go area' for trams in the heart of London would have been eliminated, enabling a comprehensive network of tramlines to be constructed with all the 'through-running' capabilities that this would have ensured.

With regard to the veto powers enjoyed by local authorities and 'frontagers', the Royal Commission took the view that such powers had greatly inhibited the development of a general system of tramways operating for the benefit of the public at large. Its recommendation was that they be abolished [25].

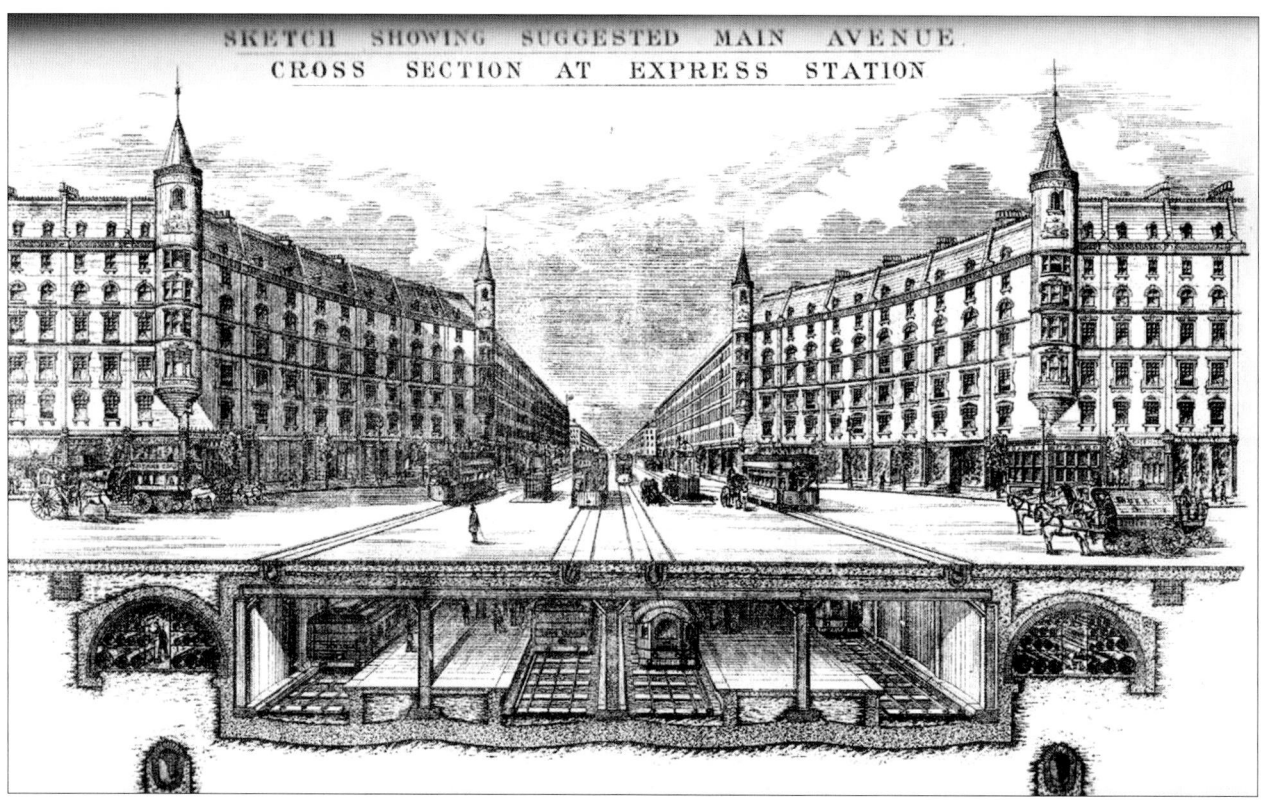

Sketch from the Royal Commission Report showing a cross section through one of the proposed Main Avenues with trams and road traffic on the surface and railway lines beneath.
(HMSO)

It also proposed that any local authority should have a preferential right to construct tramways within its district, subject to approval being granted by a Traffic Board and by Parliament.

The Commission examined the various factors contributing to London's traffic congestion, which included irregular street widths and the lack of street planning. It concluded that the chief cause of the problem stemmed from the absence of any one municipal or other authority exercising jurisdiction over the entire area together with the power and resources to tackle the various issues. Unless the problem was dealt with, it predicted, the life and growth of the capital would be strangled by the choking of its arteries with traffic.

Having persuasively identified the need for a permanent body responsible for dealing with questions of locomotion in the Greater London area, however, the Royal Commission's proposed remedy was disappointingly timid. Despite being urged by some witnesses to amalgamate the various tramways in the Greater London area and place them under the operational control of the LCC, it declined to express a view on this issue one way or the other. It proposed, instead, to shelve the whole topic of municipal ownership and operation of London's tramways to be the subject of a separate investigation.

Likewise, despite the obvious connections between transport issues on the one hand, and housing and other related social policy issues for which local authorities were responsible on the other, the Commission did not feel it appropriate to either extend the territory of the LCC or to increase its powers to take on such a role. It cited the friction this could spark with other local authorities in the area [26], and proposed instead the creation of a small permanent ad hoc traffic board, with a largely administrative and regulatory remit. This was to include the examination and revision of public or private Bills relating to transport in London, and a range of detailed, largely co-ordinating, functions. This was another missed opportunity which also failed to anticipate the potential impact of new forms of public transport such as motor buses, then beginning to appear in small numbers on the streets of London.

Just fifteen years after the Royal Commission reported, an influx of motor buses and other commercial vehicles, many of them decommissioned following the end of the First World War, began to clog the streets of the capital. This reignited the debate about the need for a coherent and co-ordinated mechanism for the more rational planning of London's transport infrastructure and culminated in the appointment of an Advisory Committee on London Traffic to make recommendations for the easing of the capital's traffic congestion.

In its report, published on 1 January 1920, the Advisory Committee advocated the creation of a London Traffic Authority with powers to control the whole of greater London's public transport services, together with all other forms of traffic. A similar recommendation was made at around the same time by the House of Commons Select Committee on Transport (1919). However, it was not until 1924, when a further influx of small-scale independent bus operators brought about a serious increase in traffic congestion, that the first tentative steps were taken to systematically regulate the operation of public transport services in London, instead of relying on the effect of market forces.

The impact of the 'pirate' bus operators, as they were known, was felt most keenly by London's tramways; the three company-owned systems responded by proposing to reduce the wages of their tram crews. This provoked a major strike which quickly extended to the crews of the established bus operators and also London's underground rail services. The Government of the day, a minority Labour administration led by Ramsay MacDonald, successfully resolved the issue by agreeing to the striking tram drivers' demands for higher pay. It also addressed the concerns of their employers relating to the unfair competition they were facing.

Under the terms of the resulting London Traffic Act of 1924, the Minister of Transport was given the power to designate certain streets as 'congested', which effectively banned other operators from plying for hire on those routes [27]. As most of the affected streets were major tram routes, this provided an interim measure of protection for London's tramways, pending the adoption of a more permanent solution: the wholesale reorganisation of London's public transport, brought about by the London Passenger Transport Act of 1933.

Failure to reform the regulatory framework governing London's public transport infrastructure during the critical period of expansion around the turn of the twentieth century was undoubtedly the biggest single factor precluding the development of a coherent and comprehensive transport system.

If such a system had been developed, London's tramways might have been able to play a much more pivotal role than they did. The consequences of failure were entirely predictable: a public transport free-for-all in which rival transport modes sought first to establish and exploit the most profitable routes and then to engage with one another in a ruthless quest for market dominance. This was a contest in which the LCC and London's tramways in general were always likely to come off worse, regardless of the merits of individual tramcars such as LCC No. 1.

Free-for-all for London tramways' rivals

As we have seen, London's tramway operators were not only hampered by a restrictive regulatory framework but were also subjected to a virtual 'banning order' excluding tramcars from much of the city centre and heart of the city's commercial and cultural district. This 'tram-free' vacuum at the heart of London encompassed an area approximately two miles by six miles in extent [28]. Rival bus operators, who were not constrained by either impediment, were quick to take advantage of this by exploiting the public transport vacuum in central London.

Initially, this conferred an immediate benefit on London's existing horse bus operators [29]. They were effectively granted a stay of execution in central London, enabling them to retain an anachronistic and inefficient mode of transport which was rapidly being displaced in other parts of the country, firstly by horse-drawn tramcars but increasingly, by the turn of the century, by expanding electric tramway systems.

Horse buses had first appeared on the streets of London in 1829. In 1855 an Anglo-French company, the London General Omnibus Company (LGOC), was founded (under its original name Compagnie Générale des Omnibus de Londres) with a view to purchasing and amalgamating London's many horse bus operators, most of which were ailing as a result of excessive competition. By 1863, the Company's omnibuses accounted for 85% of the total, and horse bus numbers peaked at around 3,700 by the turn of the century.

Even as late as the mid-1890s, London's horse buses were still carrying more passengers - around 300 million – than its horse tramways were, at around 280 million [30]. Elsewhere in the country, their numbers were rapidly eclipsed during the following decade by the rapid expansion of electric tramways; in Liverpool, for example, horse buses disappeared from the streets within seven years of the arrival of electric trams. In London, by contrast, they held on much longer, with the last LGOC horse bus lingering until October 1911, while the last privately operated horse bus lasted until August 1914 [31].

A Putney to Wimbledon horse-drawn bus photographed at the bottom of Putney Hill on 20 July 1912, long after such vehicles had disappeared from most provincial towns and cities.
(© TfL from the London Transport Museum Collection)

However, it was the petrol-engined motor bus which was to benefit most from the absence of electric tramway competition in central London during the early years of the twentieth century. The first motor buses appeared in 1899 but, with their solid chain-driven wheels and standard horse bus bodies, these primitive pioneers struggled to establish themselves in competition even with the increasingly outmoded horse bus. They would have stood little chance in competition with a contemporary electric tramcar. As it was, several early motor bus operators went into liquidation [32].

Breakdowns were common and accidents rife until tighter regulations were introduced following a serious accident in 1906. Effectively shielded from tramway competition, however, the design, efficiency and reliability of motor buses gradually began to improve. The LGOC, having initially sought to take advantage of its quasi monopoly by resisting the introduction of the new technology, now saw its dominance challenged by a variety of new operators. In 1902, growing competition forced the company to introduce its own motor bus services.

An increasingly rapid displacement of horse buses by petrol-engined buses followed, with motorised bus numbers quadrupling from 250 in 1906 to 1,000 by 1908 [33]. In 1909 the LGOC began manufacturing petrol-engined buses to its own design and, a year later, it introduced the first mass-produced double decker bus, known as the B-type, which revolutionised the economics of public transport operation in the capital. By 1913, 2,500 of these vehicles had entered service. With a 34-seat capacity and an official top speed of 16 miles per hour, the B-type easily out-performed the few remaining horse buses and contributed to their ultimate demise.

As a newer form of public transport, motor buses gradually gained in popularity at the expense of the more old-fashioned electric tramcar. This process continued as a result of the buses' shorter operational lifespan and consequent much more frequent upgrades, each of which incorporated the latest technological developments. As it was much easier, and cheaper, for bus operators to introduce new passenger services on routes of their choosing, they enjoyed various competitive advantages that were denied to tramway operators.

One of these advantages was their exclusive right to ply for hire in those streets of central London which were effectively barred to tramcars. Also benefiting hugely from the exclusion of tramcars was the

A German Daimler double-deck open-top motor bus with steel tyred wheels and a white 26-seat horse bus body. The bus began regular service on 9 Oct 1899 on a route between Kennington and Victoria Bridge, run by the Motor Traction Co. Ltd. The route was changed to Kennington - Oxford Circus in 1900 but the buses were withdrawn in December 1900.
(© TfL from the London Transport Museum Collection)

One of the first LGOC motor buses, with Swiss-built 'Orion' chassis and horse bus body, operating the Hammersmith to Piccadilly Circus route via Kensington between December 1904 and 1906.
(© TfL from the London Transport Museum Collection)

LGOC B-type bus no. B186 running on service 1 between Cricklewood and Elephant and Castle between 1910 and 1914. (© TfL from the London Transport Museum Collection)

network of underground railway lines. The first of these, the initially steam-hauled Metropolitan Railway, was opened in 1863. The electrically operated City and South London Railway, first of the deep level tube railway lines, was opened in December 1890.

During the late nineteenth and early twentieth centuries, unbridled underground railway construction by various promoters resulted in the creation of a somewhat haphazard network of competing and frequently overlapping routes. The various schemes were funded by the issue of shares, their primary motivation being the pursuit of profit, rather than the development of a comprehensive network that would most effectively meet the public's transport needs.

As in the case of tramways, Parliamentary approval was required for such schemes. While approval was by no means guaranteed, duplication of an existing service was not necessarily a bar as competition in transport provision was generally supported by legislators at the time. High density routes were frequently favoured even though the network might have benefited from a more even spread of services, avoiding unnecessary duplication of routes.

Charles Tyson Yerkes.
(Charles T Yerkes Collection,
New York, 1904)

Most importantly, there were no local authority or other vetoes and, consequently, no 'no-go' areas for underground and tube railway promoters. Even so, the promotion and construction of such lines was a costly, uncertain and risky undertaking, the successful accomplishment of which owed much to the intervention, in 1899, of an entrepreneurial American financier and former Chicago tramway developer, Charles Tyson Yerkes. He not only succeeded in raising the colossal sums of money required from American financial syndicates, but also ensured that most of the risks were borne by the investors, few of whom in the end secured the rewards they were anticipating [34].

Yerkes, on the other hand, was able to use the funding he raised to acquire development rights in previously moribund schemes, and to develop, electrify and extend other weak or failing ventures. In this way he built up a formidable portfolio of underground railway initiatives. In 1902, these were placed under the control of a newly established holding company, the Underground Electric Railways Company of London (or UERL).

Although Yerkes himself died in 1905, the UERL was to play a key role in the development and organisation of London's public transport infrastructure over the next three decades. By this stage, however, the preceding era of virtually unbridled public transport competition had rendered the financial prospects of the various tram, bus and tube operators in London increasingly precarious.

The privately owned public transport operators responded by seeking to alleviate the competitive pressures through a process of consolidation and empire-building. As part of a municipal enterprise, such options were clearly not available for LCC Tramways, although it was not immune from the consequences of such activities. The extent to which the LCC or its Tramways Department may nevertheless have contributed to the latter's own ultimate downfall will be examined in a later section.

Consolidation and empire-building by London tramways' rivals

By 1906-7, competitive pressures were beginning to take a toll on the growing number of bus operators. This was reflected in the above-mentioned quadrupling of the number of motor buses operating on the capital's streets, and a consequential major increase in congestion, particularly in central London.

Many smaller operators fell by the wayside while the larger ones responded by seeking to consolidate through amalgamation where possible, or, alternatively, by drawing up fare arrangements in order to reduce the threat of competition from any remaining rivals. One of the most significant amalgamations took place in 1908 when the LGOC bought out two of its main rivals, the London Road Car Company and the Vanguard Company, resulting in a combined fleet of 899 vehicles, a quarter of which it then scrapped [35].

The LGOC also took advantage of regulations introduced by the Metropolitan Police to limit the permissible weight of buses, in developing the above-mentioned design of its innovative B-type bus [36]. This was built in-house rather than relying on existing bus manufacturers and, being lighter than its competitors and therefore quicker and cheaper to operate, it thereby enhanced the LGOC's competitive edge. So successful was this strategy that it was able to acquire one of its remaining rivals, the London Motor Omnibus Company, in March 1911, leaving only one significant independent operator – Thomas Tilling. Nevertheless, even the latter was obliged to enter an agreement with the LGOC that limited the number of buses it operated [37].

However, this apparent victory unleashed a bout of retaliatory action by other enterprises who felt threatened by the LGOC's increasingly dominant market position. These included the major independent bus manufacturers, notably Daimler and Leyland, who found themselves effectively excluded from the lucrative London bus market. Also feeling threatened by the aggressive competition from LGOC's buses on its main tram routes was Metropolitan Electric Tramways (MET), a company-owned system.

Leyland responded by redoubling its efforts to sell buses in those outer London districts where the LGOC was much less dominant; and MET retaliated by setting up its own bus company and ordering 350 Daimler buses in a bid to break the LGOC's stranglehold on bus services in central London. Both moves threatened to revive the unbridled competition and over-supply that the LGOC had been seeking for years to vanquish.

Meanwhile, similar competitive pressures were also causing problems for the various London underground operators, including the recently formed UERL. A particular threat to the latter had emerged in 1900, with the appearance of a potentially powerful rival operator backed by another major American finance group, J P Morgan, which had been a primary financier for US railroad interests during the nineteenth century [38].

The Morgan group's plans for London took the form of a proposed 38-mile tube line linking Hammersmith in the west with the north-eastern suburbs of Southgate and Waltham Abbey. The proposal relied on a prospective partnership with a company-owned tramway, London United Tramways (LUT), which was interested in developing further feeder links for its tramway network based in west London. It had already benefited from a similar arrangement with the Central London Railway.

By the summer of 1902, the future looked bright for the proposed scheme. LUT had agreed to provide three-eighths of the requisite funding; the only remaining obstacle was to secure Parliamentary approval, the prospects for which appeared promising. At this point, however, C T Yerkes and his business partner, American-born financier Edgar Speyer, intervened in support of their own rival proposal and, taking advantage of the summer Parliamentary recess, scuppered the whole project [39].

Exploiting the fact that LUT remained financially independent of the Morgan group, Yerkes and Speyer persuaded LUT's chairman to sell his shares to them. They then withdrew the company's co-operation from the Morgan syndicate's rival underground scheme, causing it to collapse before Parliament had the chance to choose between the two schemes.

While this eliminated one potential additional rival, however, other competitive pressures intensified, not only from existing underground operators but also from motor buses; there was even a spirited fight-back by rival tramway operators.

The Yerkes syndicate succeeded – with the help of a massive accumulation of capital – in completing its underground lines with remarkable speed. Unfortunately, the costs were higher than anticipated, partly as a result of new regulations, requiring the fireproofing of trains, stations and escape routes from underground tunnels, which were introduced after a disastrous fire on the Paris Métro in 1903. The need for this had been highlighted in December 1901 when a fire at Dingle underground station on the Liverpool Overhead Railway had resulted in the death of six people [40]. Meanwhile, revenues were suffering as a result of excessive competition, and interest payments on the mountain of accumulated debt were becoming insupportable.

A low point was reached in 1906 when, faced with equipment failure, disappointing passenger numbers and unsustainable revenues, the UERL contacted the LCC and announced that it was prepared to capitulate by selling its holdings to the Council. This offer was rejected by the controlling Progressive Group, however, forcing the UERL to consider alternative options [41].

Aftermath of the Paris Metro fire at Couronnes Station on 10 August 1903. (Postcards of Paris)

In the end, salvation was secured through a process of collaboration, consolidation and financial and organisational restructuring. An early attempt was made to manage the competitive pressures by brokering an agreement between bus and underground operators to fix fares and share traffic. This foundered in 1907; but a more limited arrangement between the underground companies themselves did result in closer operational co-operation, including improved signalling and joint marketing ventures.

Even this was not sufficient to stave off financial collapse for the UERL, however. Disaster was only narrowly averted by means of a substantial cash injection from Edgar Speyer in December 1907, followed by a major financial and organisational restructuring in 1908. This laid the foundations for a revival in the UERL's fortunes, significantly reinforced in 1911 when it managed to acquire its most potent rival, the LGOC, on advantageous terms. The takeover resulted in the formation of an overwhelmingly powerful Combine uniting an extensive range of underground, bus and tramcar operators. From this point on, the Combine was in a position to either absorb remaining independent bus and underground companies or, effectively, dictate terms in agreements reached with them.

This gradual reversal of fortunes also dramatically altered the balance of power in the ongoing struggle between the Combine and the LCC, which found itself in an increasingly subordinate position as time went on. Indeed, the formation of the Combine in many respects foreshadowed the subsequent creation of the London Passenger Transport Board (LPTB), resulting in the demise of the LCC as an independent tramway operator. A consequence of this was the blighting of any realistic prospect that LCC No. 1 might have helped to herald a revival of London's tramway network.

LCC and its tramway: hapless victims or hopeless strategists?

So far, the emphasis has been on the external challenges – commercial, political and regulatory – that confronted the LCC and handicapped its tramway's ability to fend off competitive pressures from motor buses and electric underground rail services. In this section, the focus switches to an assessment of the extent to which strategic or operational failings on the part of the LCC and its tramway might themselves have contributed to the latter's demise.

Various failings can be identified, for which responsibility rests unequivocally with the Council and its tramway. One problem stemmed from the fact that, from the outset, the ruling Progressive Group within the Council viewed the potential benefits of tramway operation not as an end in itself, or even as a valuable source of revenue, but as a means of furthering other important social policy objectives in the related fields of housing and employment. The Progressives on the Council were enthusiastic proponents of the doctrine of municipalism (see Chapter 1), in stark contrast to their (Conservative) Moderate opponents, who strongly disapproved of such costly municipal projects [42].

By operating its own tramways with affordable fares, the Progressives reasoned that people living in overcrowded districts in the centre of London would be able to move to more spacious accommodation in the suburbs. This was facilitated by the availability of cheap workmen's fares that the LCC (unlike its rival bus operators) undertook to provide, despite the considerable costs that this policy incurred [43]. At the same time, the Council was committed to paying decent wages and providing acceptable working conditions for its tram crews.

When it came to the practical matters of assuming control of horse-drawn tramways, extending them and switching to the much more efficient electric mode of operation, however, the Council was much less single-minded in its approach [44]. An early strategic error was a refusal to grant permission in 1895 to London United Tramways to erect overhead wires that would have provided tramway services between Hammersmith and Shepherd's Bush [45].

Another early obstacle was the Council's initial acquisition of the London Street Tramways Company, which sparked off a lengthy legal battle over the amount of compensation the latter was entitled to in return for relinquishing control of its four and a half miles of track. By the time this issue had been resolved, also in 1895, local elections had weakened the control of the tram-supporting Progressive Group, thereby strengthening the hand of their more hostile Moderate opponents.

At around this time, the LCC received an intriguing offer from a commercial enterprise called the County of London Tramways Syndicate, which appeared to have the backing of American capital and technology. The syndicate proposed that it should acquire control over all the existing horse tramway companies operating in the central London area - both inside and beyond the LCC's jurisdiction - convert them to electric operation, and operate them on behalf of the LCC, who would own the tracks, in return for a substantial share of the profits.

This arrangement would clearly not readily have been compatible with the Council's goal of using the tramway as an instrument of broader social policy, but the proposal was nevertheless only narrowly rejected by the Highways Committee. Had it succeeded, it could well have speeded up the development of an electric tramway in London by six or seven years [46]. It might even have resulted in the development of a more comprehensive tramway network furnishing a stronger challenge to the increasing dominance of rival bus and underground transport operators during the early part of the twentieth century.

As it was, the LCC struggled to acquire control of the various horse-drawn tramways in its area and lacked the technical competence needed to convert them speedily to electric traction, even after it gained Parliamentary approval to operate trams on its own account in 1896. Consequently, it was not until 1903 that the Council finally opened its first electric tramline, from Westminster to Tooting; and the process was not completed until 1913 [47].

Prior to this, faced with a disjointed tramway network, implacable opposition to electric tramways from central London boroughs and Parliament's refusal to sanction tramway construction on any of the Thames bridges, the LCC initially responded by running its own horse buses in a bid to connect its tram routes to the north and south of the impregnable central tramway exclusion zone [48]. However, this elicited a successful legal challenge by commercial omnibus companies, which forced the LCC to sell its 77 buses and 500 horses in 1902.

The Council was more successful in gaining approval for the construction of a 'cut and cover' tunnel – the Kingsway Subway – connecting Bloomsbury with the Embankment, which finally enabled it to establish a physical connection between its northern and southern networks when through services finally commenced on 24 February 1906. However, a more ambitious proposal that would have resulted in a four-mile tunnel connecting Knightsbridge with the City failed to win approval from the Royal Commission on London Traffic in 1905 [49].

This slow rate of progress in developing a viable tram network for central London was in marked contrast to the development achieved during the same period by rival underground operators, who were quick to exploit the tramway vacuum by putting forward proposals for the construction of multiple new lines. Indeed, rival company-owned tramway operators LUT and MET also made better progress in developing tramway networks in London's western and north-western suburbs, respectively [50]. Perhaps sensing the competitive advantage that their underground rivals were seeking, the LCC took advantage of an investigation into London's underground railways, conducted by a Joint Parliamentary Committee in 1901 (mentioned previously), to press the case for a more co-ordinated and rational approach.

It argued for a replacement of the existing free-for-all by a more considered process involving closer supervision by Parliament, the City, the LCC and the Board of Trade, with a remit to ensure that the underground system should be designed with the best interests of London's traveling public in mind rather than the acquisitive instincts of rival sets of shareholders. This suggestion met with a favourable initial response as the Joint Select Committee Report recommended that some central authority should be made responsible for the control and supervision of all underground railway proposals [51].

It further recommended that wherever an underground railway was proposed, the relevant county councils, including the LCC, should be given the power to either construct, or assist in the construction of, the lines with a view to encouraging their extension into thinly populated districts; the object being to relieve congestion elsewhere, even where this might not be warranted on purely commercial grounds. Unfortunately, both recommendations fell on deaf ears and the largely unregulated free-for-all approach to transport planning was allowed to prevail.

Subsequently, as mentioned previously, the 1905 Royal Commission also discussed the desirability of a central authority, this time to control and supervise all tramway and railway construction schemes. It agreed that such a body was indeed merited but it explicitly rejected the idea of vesting such powers in the LCC. This was done partly on the grounds of an alleged conflict of interest stemming from the latter having a vested interest in its tramway operations. The Commission assumed that this would render it incapable of acting impartially with regard to other rival operators.

Another reason put forward for the LCC's disqualification was its limited geographical remit, rendering it allegedly incapable of addressing the transport needs of local authorities outside the County boundaries. Whether or not the Royal Commission's concerns on either score were well-founded, its own recommendation – for a panel of experts known as the London Traffic Board - failed to gain traction and was simply ignored.

The LCC's biggest strategic error, however, was its failure to capitalise on the brief moment of vulnerability experienced by its principal rival, the UERL, in the period 1906-1908 when it encountered a combination of equipment failure, disappointing passenger numbers and unsustainable revenues, which all came to a head shortly after Yerkes's death in December 1905.

As described previously, faced with an escalating financial crisis the initial response of the American financier Sir Edgar Speyer (Yerkes's successor at UERL) was to approach T McKinnon Wood, the leader of the Progressive group on the LCC, with a proposal involving the ultimate sale of the entire UERL enterprise to the LCC in return for a short-term emergency injection of £5 million. That was the sum needed to stave off the imminent threat of bankruptcy proceedings at the hands of disgruntled investors. If it had been accepted, the proposal would potentially have handed control of its arch-rivals, the Underground Group, to the LCC, possibly enabling the latter to develop a strategic transport policy for London in which its tramways might have played a more pivotal role.

The offer was declined, however, and the opportunity was completely lost when the Progressives lost control of the LCC in the elections of 1907. Power passed to the Conservative-supporting former Moderate Party, which in 1906 had changed its name to the Municipal Reform Party. This party, which was not at all interested in expanding the Council's role in co-ordinating London's transport system, was to retain control until 1934. A London Traffic Board was established in 1907 but this had a relatively restricted remit and limited powers - it was responsible for collecting information and statistics relating to traffic and transport problems in London and for preparing an annual report for Parliament. It was disbanded in 1916 [52].

In addition to these strategic errors, the prospects for LCC's tramways were further undermined in subsequent years as a result of ongoing operational failings. For example, even after its tramway operations were up and running, the Council's continued attachment to its low fares/high wage policies resulted in an escalating revenue shortfall as the number of workmen's tickets issued rose from 10 million in 1904 to 42 million in 1910 and 77 million by 1913-14, contributing to a deficit of £65,000 by 1910-11 [53].

Although a shortfall on its revenue account could be recouped, at least in the short term, by looking to the ratepayers for support, such a policy was not sustainable in the long-term without incurring electoral consequences. It also led to a difficulty in contemplating capital investments when either rolling stock or infrastructure needed replacing.

Quite apart from the growing financial burden that such a policy imposed on the council's ratepayers, it also contributed, indirectly, to a negative perception against tramcars in the minds of other sections of the travelling public, because of their predominant association with workers' transport.

Buses did not suffer from such a negative social stigma partly because they were not subject to the same legislative obligations to offer reduced price workmen's tickets. Quite apart from their continuing technological advances, therefore, such a policy further increased their competitive advantage at the expense of LCC's trams.

Conclusion

In Russian folklore, and also that of many other cultures, the blue bird was a symbol of hope. For a time, LCC No. 1, nicknamed 'Bluebird', carried the hopes of LCC's tramway department that it might help to spearhead a revival for the London tramcar in the face of the remorseless competition from rival modes of transport.

However, it was Bluebird's misfortune to have been conceived at the onset of a period of severe economic dislocation and political turbulence, coinciding with a major shift in the direction of public transport policy. The ultimate result was the demise of London's tramcars and their gradual replacement by trolleybuses and, ultimately, diesel-engined buses. This demise was not simply a matter of unfortunate timing and geographical constraints but was, as we have seen, the culmination of a variety of factors which conspired to preclude the development of a comprehensive and efficient tramway network in London that would have been capable of competing on an equal basis with rival transport operators.

The failure to develop such a network can be attributed to a variety of causes. They include the actions of hostile Parliamentary representatives who, for decades, were unwilling to countenance the introduction of a central authority capable of formulating a more rational and coherent transport policy, and with a duty to ensure that public transport proposals would only be supported if they could be shown to be capable of best meeting the needs of London's travelling public. They were also unwilling even to contemplate a reform of the byzantine regulatory regime that subjugated the meeting of London's transport needs to the vested interests and parochial preferences of rival sets of borough councillors or the occupiers of properties fronting on to proposed tramway routes.

Another cause lay in the machinations and manoeuvrings of rival transport operators, many of which were backed by wealthy American financial conglomerates. Their primary motivation was to elicit the speculative interests of their shareholders, whose investment was required in order to provide the capital necessary for massive infrastructural developments rather than meeting the needs of the capital's travelling public.

The final explanation for the failure to develop a comprehensive tramway network for London stems from a series of operational shortcomings and strategic errors for which the LCC and its tramway were themselves responsible.

Epilogue

Just under a century after work began on the construction of LCC No. 1, history seemed to be about to repeat itself when a global pandemic unleashed a period of unprecedentedly severe social and economic upheaval with profound and deep-seated effects on people's lives, livelihoods and also morale.

The Tramway Museum at Crich was one of countless organisations to be affected by a near total country-wide lockdown that commenced in March 2020. This forced a temporary cessation to the ongoing restoration of LCC No. 1, in an uncanny echo of the three-month suspension to its original construction programme – described more fully in Chapter 5 - that resulted from the Great Stock Market Crash of 1931.

At the time of writing, in the summer of 2021, the tentative reopening of the Tramway Museum and resumption of restoration work on the tramcar elicited the earnest hope that Bluebird would soon usher in a more hopeful, prosperous and sustainable era for the Tramway Museum, its visitors and also the wider world.

Chapter 4
The need for London County Council No. 1
Richard Sykes

Changes and limitations

In the preceding chapters we have seen how, during the 1920s, social and technical changes and regulatory constraints on the London County Council (LCC) for providing tramway services led to the need for the production of an experimental tramcar which was to become London County Council Tramways No. 1 (LCC No.1).

The tramways within the control of the LCC provided services for some 636 million passengers during the year ending March 31 1919. This figure increased to 685 million in 1919/1920 and to over 689 million in the following year, to the end of March 1921 [1]. Services continued to expand: for example, extensions along Eltham Road were approved in November 1920 [2], and proposals for the 1922 session of schemes included Shepherd's Bush to Marble Arch and Victoria, and Grove Park Station to Lee Green [3].

London County Council Tramways (LCCT) continually sought to encourage patronage of the tramways by improving services and conditions for the travelling public. They implemented the LCC's low fares policy and also provided lower workmen's fares; the latter had been imposed on some of the original horse car companies, but not on the LCC, by the Tramways Act of 1870 [4] [5]. A further lowering of fares, an increase in the distance between fare stages, the introduction of a new 'midday' fare, and other fare structure changes were to produce high patronage of the tramcars by 1923 [6].

However, as we have seen, their services had to terminate at boundaries of the City of London and the West End, as the Corporation of the City of London possessed a veto which was invariably used [7] [8]. As early as 1905, the report of The Royal Commission on Means of Locomotion and Transport in London had recognised that tramway mileage was insufficient and the lack of through routes caused great inconvenience [9].

Procedures and personnel within the LCCT

The LCC Tramways Department management reported to the London County Council Highways Committee, as the body responsible for the tramways, at regular - generally weekly or fortnightly - meetings. Approval given by the LCC Highways Committee for tramway projects requiring finance could then be subject to further scrutiny by the LCC Finance Committee before final approval.

The position of LCC Tramways Chief Officer or General Manager was held by Mr Aubrey Llewellyn Coventry Fell from 1903 until December 1924; by Mr Joshua Kidd Bruce from 1925 to June 1930; and by Mr Theodore Eastaway Thomas, from June 1930. In the period we are concerned with, Mr W E Ireland served as Rolling Stock Engineer until 1930, when he was succeeded by Mr G F Sinclair.

New tramcars to meet traffic demand

In 1920 a request from the LCCT for the provision of 125 new tramcars to meet the increasing traffic demand was approved. The tramcars were to be of the type already in use, known as Class E/1, with a design first introduced in 1907. New designs were stated as being for consideration, but between two and three years would elapse before any delivery, and the General Manager was anxious for new cars to be obtained as soon as possible [10]. Agreement to obtain tenders was made on 17 June 1920; delivery of 125 Class E/1 tramcars was planned to begin in September 1921 [11]. Generally the Class E/1 tramcars were claimed to be far ahead of their time. However, technical and structural improvements were not generally noticeable to the travelling public as the overall appearance was unchanged, at a time when buses were making obvious, frequent and significant advances in design and size [12].

The new tramcars were fitted with 60 horsepower motors, two per car, giving increased speed and more rapid acceleration than other tramcars in the fleet. These factors reduced running times for a journey, allowing more journeys to be made with fewer tramcars [13]. During 1923, trials with and without

1908 design of Class E/1 LCC tramcar built in 1920. (Hurst, Nelson/J H Price collection)

Class E/1 upper saloon with wood slat seating. (© TfL from the London Transport Museum Collection)

Class E/1 lower saloon with longitudinal wood seating. (© TfL from the London Transport Museum Collection)

trailers on south London services were conducted, with these high power tramcars – designated tractors - used for towing.

A week's trial was made with tractor and trailer, followed by another week with tractor only; the results were generally the same. On the Streatham and Norbury route, for example, 178,386 passengers were carried with the trailers being used, and 192,534 with tractor tramcars only. Revenue was £1,431 3s 9d and £1,521 5s 3d respectively. The General Manager concluded that the advantages of non-trailer working were increased speeds, reduced delays due to accidents and breakdowns, and covered accommodation [14] – the trailers being open-topped vehicles. Trailer operation consequently ceased in 1924 [15].

The bus as the major competitor

Even before 1920 the motor omnibus became a major competitor to the tramways in London. In much the same way, tramways had been deemed cause for improvements in suburban rail travel earlier in the century, with cited cases highlighting the need for promoting electrification of the suburban railway network [16] [17]. By 1920 the tramways were losing traffic. The number of passengers carried by LCC tramways fell by a little over a million to 688 million in the year to March 1922. After an increase to nearly 734 million passengers in the following year, numbers declined to 689 million in the year to March 1924 and to 682 million in the year to March 1925 [18].

Cutting fares resulted in some increase in passenger numbers, but this did not necessarily generate higher takings [19].

Tramways, however, continued to receive support, as can be seen from this statement by G H Hume, JP, to London Members of Parliament [20]: "It is within our knowledge that much has been said in various quarters as to the tramways being obsolete. In view of this it may be well to submit a few figures to illustrate the part which the tramways system plays in meeting the traffic requirements of London today.

Passengers Carried	1919	1920
All London Tramways	1,052,744,820	1,065,788,631
Omnibuses	861,001,965	935,946,002
Tube Railways (Combine)*	266,590,055	265,587,941
Other local railways	396,000,000	424,000,000

"It will be observed that the tramways carry the greatest number of passengers.

"During the rush hour traffic periods the tramways service of the Council is increased by 48 per cent to enable the cars to accommodate those having to travel. No evidence exists, however of any real effort on the part of the omnibus undertakings to deal with this special traffic. To illustrate this point:

	Tramways		Omnibuses	
	Seats	Passengers	Seats	Passengers
Up traffic 7-10am	16,234	14,2275	2,798	2,306
Down traffic 4-7pm	16,486	14,541	3,742	3,692
Other hours	98,044	27,384	25,094	16,343

"For the year 1920 the comparative costs of carrying each passenger was as follows

'Combine' railways*	2.20d
Omnibuses	2.09d
London County Council tramways	1.61d."

*The Underground Group of companies.
(The tramways figures in these tables include the LCC tramways and Company tramway operators in the London area.)

To overcome both the restriction to tramways entering the City of London and the competition from motor buses, the LCC applied to Parliament for powers to operate buses, to have similar benefits and possibly even to replace tramcars on some routes. Following opposition by the London General Omnibus Company (LGOC) and other bus operators, permission was refused [21] [22]. The LCC, having to continue to operate tramcars only, then sought to extend tramway services.

Motor bus traffic continued to increase, with independent bus operators beginning to offer services to the public; Arthur Partridge, the first of these to be licenced, introducing the 'Chocolate Express' in 1922. This extra competition, added to that of the LGOC buses, prompted LCC Tramways to seek protection for its own services. It petitioned for restrictions to competition from other modes of public transport on roads where tramcars ran, and gained these with the passing of the London Traffic Act of 1924; the Act limited the number of buses which could operate along designated restricted streets. In fact, most LCC tram routes became restricted streets [23] [24] [25].

Developments of the motor bus

Improvements in the design of motor buses for use in London started before 1920, when the 46-seat K-type superseded the pre-war B-type with its 34 seats. In 1923 the LGOC introduced the NS-type, with a 52-seat body, constructed on a drop frame chassis which produced a lower centre of gravity. Although this model was fitted with an open-top body, it was designed for covered upper saloons in anticipation

1923 production of NS-type bus of low chassis design with open top deck. Chassis designed to enable a top deck cover to be fitted. (© TfL from the London Transport Museum Collection)

Upper deck seating of 1923 NS-type bus. (© TfL from the London Transport Museum Collection)

Lower saloon seating of 1923 NS-type bus. (© TfL from the London Transport Museum Collection)

1926 production of NS-type bus following Metropolitan Police approval for a top deck cover as envisaged in the original design. (© TfL from the London Transport Museum Collection)

Standard ST-type AEC Regent bus of 1929 (© TfL from the London Transport Museum Collection)

Upper saloon seating of 1929 ST-type bus. (© TfL from the London Transport Museum Collection)

Lower saloon seating of 1929 ST-type bus. (© TfL from the London Transport Museum Collection)

of the removal of police restrictions on the use of top covers. From 1925 nearly all the NS-type buses were built with top covers, and many were converted from solid to pneumatic tyres [26] [27]. With dimension restrictions and to reduce weight, which was also regulated, these motor buses were developed with a bodywork of composite construction, with metal panels on an ash body frame.

Following some three-axle chassis designs, by 1929 the bus design for London had progressed to the AEC Regent chassis, with a double deck body designated type ST. Predominantly, these had Chiswick-built bodies with a full-width enclosed rear platform and a straight staircase, with seating for 29 upstairs and 20 on the lower deck [28]. The AEC Regent, together with the Leyland Titan, were to be the principal double deck bus chassis throughout the UK for more than a decade.

With all of these developments the passenger numbers that could be carried in any of these motor buses did not meet the capacity of the 74 seats provided on the Class E/1 tramcar.

The LCC's position on tramways in 1924

In July 1924 the LCC produced a report entitled *Tramways, Trolley Vehicles and Motor Omnibuses Compared* [29] [30]:

"The advantages and disadvantages of each system have been summarised by the General Manager of the Council's tramways as follows:

"*Seating accommodation* - The tramcar has normally approximately 50 per cent greater seating capacity than either of the alternative vehicles. *In wet weather* compared with the petrol omnibus the advantage of all seats being under cover is 200 per cent.

"*Speed* - The tramcar is under better control and has a higher standard of acceleration and deceleration. On the move it is capable of higher speed than either of the alternative vehicles.

"*Safety* - Being confined to the rails, the tramcar is least liable to accidents and is not liable to skidding sideways.

"*Cost of operation* - Tramcar costs least a set mile to run.

"*Wear and tear of roadway* – Practically no damage is done by the tramcar, whereas both the petrol and the trolley omnibus inflict damage out of all proportion to their contribution to the Road Fund.

"*Operation in bad weather* - Experience has proved that the tramcar can continue running in foggy or snowy weather long after the omnibus services are suspended.

"*Contribution to local rates* - Trolley and petrol omnibuses make no payments in respect of local rates for 'track'.

"*Flexibility* – The omnibus has greater flexibility, but the experience is that this advantage is least available when most required, as, for instance, in congested areas like Piccadilly Circus, the Bank and Fleet Street - the average speed is not more than 6 miles an hour.

"*General conclusions as applied to London conditions are*:- The main consideration however, must always be the financial one, and, while the tramcar has the big advantage in seating accommodation, the cost per seat mile for operation must always be considered less than that of smaller vehicles.

"It must not be forgotten that the tramcars are excluded from the profitable central London area.

"The question of the provision of trolley omnibus services or petrol omnibus services as extensions of existing tramway services is one that undoubtedly merits serious consideration where competition by another means of transport does not exist. It would appear however that short extensions, involving a change of vehicles, would not be successful in cases in which a through service is provided. However advantageous the operation of trolley omnibuses might be from the point of view of operating expenses, there are not many cases of extensions by means of trolley omnibus services that would provide such satisfactory results as extension of the tramways."

Forward thinking by the management of LCC Tramways

A criticism may be made of the LCC's lateness in appreciating the need to improve the image of its fleet of tramcars. It has been stated that the modernisation of existing tramcars did not start until 1926, and apparently then only as a result of unfavourable comparison between their cars and those of Walthamstow Corporation [31]. However, even in 1920 the Manager had stated that new designs were under consideration, although two or three years would elapse before any delivery. Some of the ideas and actions within British tramways from 1920 onwards indicate that LCC Tramways' staff were progressive in their thinking and promotion. They also maintained close contact with their counterparts in both municipal and company tramways.

In 1920 a competition for a new design of tramcar was implemented by LCC Tramways [32]:

"Competition for new design of tramcar
"The instructions for the competition were for: "The design for a new electric tramcar for use on London County Council tramways and it has been decided to invite engineers and others possessing the requisite technical knowledge and experience to submit designs for this purpose.
"The assessors of 76 proposals submitted were Mr J P Crouch, General Manager of the Metropolitan Carriage and Wagon Co., and Mr A L C Fell the LCC Tramways General Manager.
"For selection, the schemes were separated into categories, along with numbers received:

> "A Originality in design of complete car suitable for service on the Councils (sic) tramways   Nil
> "B Originality in design of complete car but unsuitable for service on the Councils (sic) tramways, including designs that do not comply with the regulations issued by the Commissioner of Police and the Ministry of Transport (Board of Trade)   Nil
> "C Designs worked out in a workmanlike manner but which are not original in design, or suitable for service in London and do not comply with the regulations issued by the Commissioner of Police and the Ministry of Transport (Board of Trade)   26
> "D Designs containing useful original and/or novel suggestions of a minor character   8
> "E Proposals suggesting original impractical features   5
> "F Designs suggesting modifications to the existing standard design of car   10
> "G Designs embodying well known features on other undertakings which are unsuitable for the LCC Tramways   1
> "H Rough sketches, incomplete particulars and specifications and entries not complying with the competition conditions   22
> "K Written matter only, unaccompanied by sketches or designs   4

> "No proposal merited placing in Category A, where the prize winning design would have appeared. Limited details were cited in the Assessors (sic) report to the Highways Committee [33]. These did however include statements that there were a dearth of proposals in respect of electrical equipment, and in a majority of cases a complete lack of electrical and mechanical knowledge was apparent. Centre entrances, reversed staircases for quick loading were mentioned even though these were not permitted by the Police and modifications to the platforms of the existing cars.
>
> "The assessors concluded that the best of the proposals fell short of the existing design in respect of safety to the public, seating capacity, stability and sound or economical constructional features."

The assessors, in fact, found the result disappointing, as nothing really new and practical had been suggested [34].

At the Tramways and Light Railways Association and the Municipal Tramways Association conference in 1923, Mr W E Ireland presented a paper on the development of the modern tramcar. Topics included weight reduction, speed increases and the use of single deck coupled tramcars [35].

Three years later, the Joint Standing Committee on Rolling Stock of the Tramways and Light Railways Association and the Municipal Tramways Association [36] concluded its work and produced a report. The LCC was represented on the Committee by Mr W E Ireland and the tramway companies' representatives included Mr C J Spencer of the Metropolitan Electric Tramways (MET).

> **"Joint Standing Committee on Rolling Stock**
>
> **"The report on Car Bodies**
> "- provided comparative data on staircase doors, vestibules, body construction, weight, entrances and exits.
> "It was concluded that; with varied conditions of traffic, standardisation was not practical.
>
> **"The report on Trucks and Equipment**
> "- Considered high-speed motors, their unsprung weight and electrical control, silent running and general noise.
> "Further work was recommended for; the motors and silent running, sandwich wheel construction.
> "It was concluded that; reasons for general noise were design and levels of maintenance." [37]

The Chairman, when presenting the Report to the Association Annual Congress in Torquay in 1926, is reported as saying: "Tramways have the advantages of a special track and a steel rail, which allows us to use, within reason, vehicles of almost any capacity. There are not the same limitations either of weight or size as is (sic) necessary with an ordinary road vehicle, and the use of electric power gives unlimited power and it can give us the most silent and most effective street passenger vehicle possible.

We have potentially a vehicle capable of handling mass traffic, and mass traffic is the feature of the traffic problem. With regard to the reduction of weights of bodies, we do not buy electricity to run dead weights about. Our primary function is to carry passengers, and the body should not be one ounce heavier than is absolutely necessary for stability and comfort. We have a great deal to learn from our 'bus friends in that direction." [38]

Of the committee having undertaken the work, the Chairman said: "They have advanced this industry considerably, not only as a result of this report, but even more as a result of the meetings with each other." [39]

In the 1920s, was the trolleybus a competitor?

At this period it was considered that trolley vehicles would complement and not compete with tramways as they could be of value for short extensions to existing tramways. Requiring only additional overhead to be erected, they would be a more economical alternative to a tramway as a first cost.

Lighter weight and less tyre wear were also claimed as technical advantages of the trolleybus over the motor bus. It was conceded that with the carrying capacity of the tramcar "...in general, it may be summed up that for dense passenger carrying in thickly populated areas the double-track tram system may still hold its own for some time." [40] [41]

However, this was to change with the proposition for replacing tramcars with trolleybuses. The use of trolleybuses rather than petrol buses was partly based on the provision of the existing supply of electricity from municipal or company power stations.

In the 1920s, experience with trolley buses was mixed. In 1925, Alfred Baker, General Manager of Birmingham Corporation Transport, was asked by Ipswich Corporation to report on suggested transport for Ipswich. He replied: "I have no hesitation in advising that it would be wise policy to abandon the present system of tramways and in place thereof substitute an up to date system of trolleybuses. I advise this not only from financial considerations but from the point of view of the comfort and convenience of the travelling public. With regard to operating expenses the cost of running motor buses is heavier than either tramways or trolley vehicles." [42]

Oldham Corporation had replaced one tram route with single deck trolleybuses in 1926, but their use was short-lived. "The failure in Oldham of the trolleybus system was due to the fact that the machines were fitted with solid tyres and as a result of running over a road paved with granite setts there was a continuance of serious complaints of vibration. It was impossible at the time to repave the road and the design of the vehicles was such that it was impossible to substitute pneumatic tyres for solids." [43]

In 1922, trials on part of the London United Tramways (LUT) system had been carried out [44] [45], and authorisation sought for conversion of the Twickenham Junction to Richmond Bridge line. This failed due to objections from Twickenham [46], but powers were gained in 1930; the first LUT trolleybus was delivered in January 1931, and the first service started in May of that year [47] [48].

Trolleybus as supplied by Railless in 1926 to the Oldham and Ashton system which operated for two years. (Tramway & Railway World)

Trolleybus supplied to Southend in 1930, used for the replacement of a tramway route. (Tramway & Railway World)

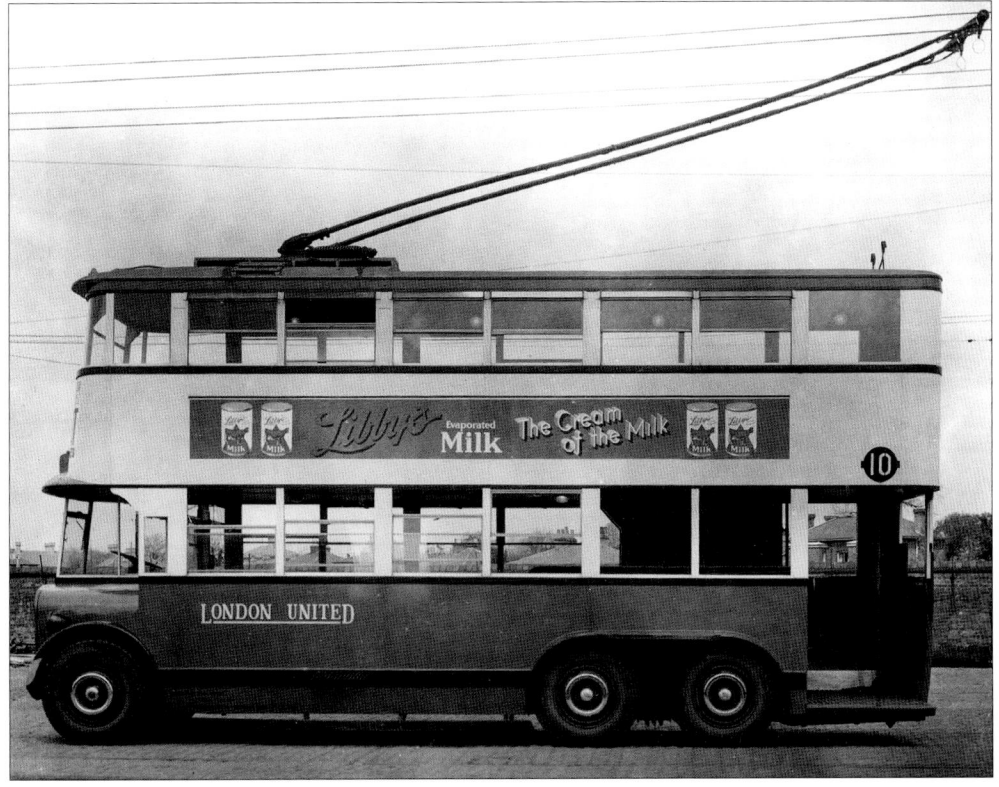

1930 trolleybus with bodywork produced by UCC Feltham for LUT. (© TfL from the London Transport Museum Collection)

London Underground Tube stock operating in 1920. (© TfL from the London Transport Museum Collection)

Interior of Tube stock operating in 1920. (© TfL from the London Transport Museum Collection)

New London Underground Tube stock built in 1930. (© TfL from the London Transport Museum Collection)

Interior of 1930-built Tube stock. (© TfL from the London Transport Museum Collection)

Underground competition and extensions

Extension of the Underground network was taking place with the Hampstead Tube, later to be the Northern Line, reaching Watford in 1922 and Edgware by 1925, and lines in the south extending to Kennington and Morden in 1926. The Piccadilly Line to Finsbury Park was to be extended to Cockfosters in 1932, directly influencing the tramway services to Manor House which would no longer act as a major interchange.

Modernisation of existing LCC tramcars

Having placed orders in 1920 for the supply of a further 125 Class E/1 tramcars, LCC Tramways then considered measures for increasing the appeal of tramcars to the travelling public.

Drawings were produced in 1922 for tramcars with 'passenger flow', for passengers to enter at the rear of the tramcar and exit at the front [49]. Other measures considered – additional ventilation, four additional lamps per saloon, and service number end plate improvements - appeared in a report to the Highways Committee on 15 November 1923 [50].

LCC Tramways' General Manager, Mr A L C Fell, had considered it impossible to fit transverse seating in the lower saloon of conduit-equipped tramcars without obstructing the floor hatches which needed to be kept free for access. In 1925, however, a scheme was produced by his successor, Mr J K Bruce, which would allow transverse reversible seats to be installed by fixing the seat frames to the steel side truss plates and carrying them on cantilevers over the floor space [51].

A number of improvement trials were being conducted on a limited basis. Authorisation for a trial of lower deck heating was reported on 24 October 1924; eight tramcars were fitted with heaters below the longitudinal seating. Mr Bruce reported on 31 March 1927 that trials had shown that a possible temperature gain above ambient could be achieved. Although consideration was given to trials on one route, Service 7, with a requirement for 18 cars to be fitted with heaters, Mr Bruce did not recommend this, partly because tramcars were then being fitted with transverse seats which increased the costs for heaters. He also considered that the heating was only of value on a few days in the year. This idea was not pursued further [52].

In 1927, uncollected fare boxes, to be used when the conductor had been unable to collect all the passengers' fares, were considered, but rejected on experience from other municipal users which indicated that costs of fitting would be greater than likely revenue.

The first LCC Class E/1 tramcar No. 1817 with modifications described as the Pullmanisation programme, including new brighter livery and improved seating. (© TfL from the London Transport Museum Collection)

The Pullmanisation programme – upper saloon with upholstered seats, but not the seat backs. (Tramway & Railway World)

The Pullmanisation programme – lower saloon with upholstered seating. (Tramway & Railway World)

Highways Committee minutes from 16 July 1925 [53] state that authority was given for five tramcars to have improvements to indicator boxes and to the plates for the service numbers.

Class E/1 tramcar number 1235 was subject to several experimental modifications. It may have been used as a 'pay as you enter' [PAYE] mockup before 1921, and by early 1924 it had been given large service numbers and cushion spring seats in the lower saloon. By 1926 transverse seating had been fitted in its lower saloon, and all the lower saloon seating had been upholstered. Its ventilation was improved, draught prevention measures were taken, and opening windows were trialled in the lower saloon. Modifications were also made to the upper deck windows to lessen obstructions to passengers' lines of sight, and protection was provided over windows to reduce water ingress. In both saloons additional lighting and ceiling handrails were installed. This tramcar also had a new form of roof. 1235 was presented to the Highways Committee to view on 21 December 1926 [54].

In the same year, modification trials were agreed by the Highways Committee on 28 April, and in May E/1 Class tramcar 1817 was selected to be fitted with transverse seats and cork carpet flooring in the lower saloon. It was painted in a new red livery and inspected by the Highways Committee on 17 June.

Then, on 17 July, Mr Bruce was authorised to modernise 100 E/1 tramcars. The modifications consisted of cushioned transverse double and single seats in the lower saloon, with grey patterned moquette-covered spring seating from G D Peters; four additional lights in the lower saloon; NUMA air-operated bells; ceiling hand rails and white ceilings. The livery was also modified from a darker Midland Red and cream to a brighter crimson red and cream livery. No change was made to upstairs seating and the modernisation omitted the cork flooring.

This modest updating was advertised to the public as a 'Pullman' car. The work on the first 100 E/1 tramcars was done at the recently expanded Charlton Repair Works during 1926/27. Modernisation of a further 250 cars was authorised on 16 February 1927; this work took place in 1927/28 [55]. With accrued savings within the modernisation programme, it was agreed on 1 November 1927 that the upper deck seats of these 250 tramcars should be upholstered; and when, on 6 March 1928, agreement was given for another 500 tramcars to be modernised, upholstery for the upper deck seats was included from the start [56].

Further work on 549 tramcars, of E, E/1 and M classes, was approved on 29 January 1929. By late 1932 they had been equipped with upholstered seats and incidental improvements had been made, but still only the lower saloon seats were supplied with upholstered backrests. On the upper decks, plain wooden backrests had been retained. Capital was requested for purchasing upholstered back rests, and the LCC Highways Committee noted: "In view of the demand for increased comfort on public transport vehicles it is desirable that the back rests of the upper deck seats should be upholstered on a number of these tramcars." [57]

As may be seen from some of the dates relating to LCC Tramways' trials on improvements and, subsequently, experimental new tramcars, the times between some trials and implementation on a large scale could be short, or could even overlap.

Fares and publicity

Along with promoting fares, posters advertising locations and events which could be reached by LCC tramways were also used to demonstrate some of the ongoing developments to increase benefits for passengers, comparing the tramcars and their services favourably against the motor bus competition. These illustrated colour posters highlighted the value of covered top decks in the rain, the quality of lighting for reading whilst travelling, the speed of services and the improvement in comfort introduced with the Pullmanisation programme [58]. The advertising itself was having to compete with the bus and underground network's colourful promotions, although even in 1930 the LGOC's advertising still stylised buses as being open top. The spread of the Underground railways was to be seen in their posters mapping the new extensions.

Modernisation developments by the London tramway companies with new tramcars

In modernisation experiments with completely new vehicles the LCC lagged behind the tramway companies in London who were members of the bus and underground combine, London United Tramways and Metropolitan Electric Tramways.

In 1925 C J Spencer, the MET Manager, had prepared drawings of a tramcar with various improvements which had been submitted to the engineers of the London General Omnibus Company for their criticism [59] [60]. The design was influenced by the work of the Tramways and Light Railways Association and the Municipal Tramways Association's Joint Standing Committee on Rolling Stock, of which, as noted previously, Mr C J Spencer was the Chairman [61]. It was referred to the LGOC engineers for an opinion, since they had developed the NS type motor bus in 1923, and the same construction principles could, it was thought, be applied to tramcar construction [62].

The result was an Underground Group decision to build two experimental tramcars. One of these, number 318, built by the MET to their own design at Hendon works, became known as 'Bluebell' because of its blue livery. The other, number 319 – subsequently LUT 350 - was built by the LGOC at Chiswick, and became known as 'Poppy'. Both were completed in 1927 [63].

With experience in traffic and from passenger's opinions on each of these two experimental vehicles, in April 1928 Lord Ashfield gave instructions to W S Graff-Baker to produce designs for experimental tramcars suitable for use on the MET and LUT. These new experimental tramcars were built by the Union Construction and Finance Co. Ltd in Feltham. The results of the research into the construction of tramcars were first seen in March 1929 when a new car was shown to the press. This was MET 320.

The associated publicity claimed that the tramcar would provide maximum comfort, speedy and silent running, suitability for dealing with mass traffic by means of a rear entrance, front exit and capacious enclosed platforms, and an especially pleasing exterior [64].

MET experimental tramcar 318, as built.
(© TfL from the London Transport Museum Collection)

MET experimental tramcar 318, upper saloon seating. (© TfL from the London Transport Museum Collection)

MET experimental tramcar 318, lower saloon seating. (© TfL from the London Transport Museum Collection)

Two further experimental tramcars were to follow in 1929, as MET 330 and MET 331 [65].

The experience gained in this comprehensive programme with its substantial financial resource for the experimental tramcars and their trials resulted in defining the final design for a production series of tramcar.

Early in 1930, an enquiry by the MET for 100 new tramcars was sent to the Union Construction and Finance Co. Ltd. The purchase authorisation was approved on 8 July 1930 and the order made public in August. Delivery started in December 1930 [66].

This production series, to be known as UCC or Feltham tramcars, was constructed with steel frames and steel sheet panels [67]. Of the order for 100 tramcars, 46 were delivered to the London United system and 54 to the Metropolitan system. The first eight UCC cars went into service on Monday 5 January 1931.

MET experimental tramcar 319. Subsequently numbered 350 in the London United tramcar fleet. (H Nicol/National Tramway Museum)

MET experimental tramcar 319, lower saloon seating. (© TfL from the London Transport Museum Collection)

MET experimental tramcar 319, upper saloon seating. The Chiswick-built bodywork incorporated components used for the NS type bus. (© TfL from the London Transport Museum Collection)

1926 production of NS type bus with covered upper deck, upholstered seating, and bodywork components which were used for experimental tramcar 319. (© TfL from the London Transport Museum Collection)

MET experimental tramcar 320. (© TfL from the London Transport Museum Collection)

MET experimental tramcar 320, upper saloon trial seating. (© TfL from the London Transport Museum Collection)

MET experimental tramcar 320, lower saloon trial seating. (© TfL from the London Transport Museum Collection)

MET experimental tramcar 330. (© TfL from the London Transport Museum Collection)

MET experimental tramcar 330, upper saloon. (© TfL from the London Transport Museum Collection)

Lower deck entrance of 330 showing staircase reversed to accommodate pay as you enter (PAYE) trial arrangements. (© TfL from the London Transport Museum Collection)

MET experimental tramcar 331. (© TfL from the London Transport Museum Collection)

MET experimental tramcar 331, upper saloon. (R Sykes)

MET experimental tramcar 331, lower saloon. (R Sykes)

1931 tramcar of which 100 were produced by Union Construction Company (UCC) Feltham for the MET and LUT. (© TfL from the London Transport Museum Collection)

MET Feltham, upper saloon seating. (© TfL from the London Transport Museum Collection)

MET Feltham, lower saloon seating. (© TfL from the London Transport Museum Collection)

LUT trolleybus, upper saloon with similar design to UCC tramcars. (© TfL from the London Transport Museum Collection)

LCC Tramways' replacement programme and modernisation with new tramcars

In a report to the Highways Committee on 23 November 1927, the need to develop new tramcars for replacing C class tramcars used on hills was highlighted; and on 20 December £6,500 was approved for the construction of two tramcars [68]. Tenders for parts for the tramcars, which were to become numbers 1852 and 1853, were agreed in April 1928; they would be constructed by staff at the Central Repair Depot [69].

The two completed tramcars were presented to the Highways Committee at Stangate on 26 February 1929.

The tramcars were Class HR/1 number 1852 and Class HR/2 number 1853. The HR denoted Hilly Route classification and both tramcars had four motors, with a motor for each axle within two bogies with equal diameter wheels.

In March of that year, the LCC Tramways Department publicised the building of these two experimental Pullman tramcars for routes with severe gradients. In the technical press the bodies were described thus: "... commodiousness and comfort surpass the Pullman cars. On both decks the general planning of the seats is on similar general lines to that in the Pullman cars. The seats are of more luxurious types than those at present used, more space being allotted for each person, while the whole surface of each seat is upholstered over flexible springs and padded in such a manner that the passengers are entirely protected from contact with any structural parts of individual seats. On the upper deck the seats are constructed on lines similar to those in the lower saloon but embody, as compared with the existing cars, the appreciable extra comfort of fully upholstered and spring supported back rests. For the covering fabrics of the seats in both saloons materials of similar design and finish to those used on the Pullman cars are utilised, thus securing standardisation. Moquette velvet is adopted for the main saloon, and red leather cloth on the upper deck.

"The construction of the upper deck framing is of a wholly new and interesting type, consisting almost entirely of interchangeable metal parts. These parts are so arranged as to secure, within the same external measurement of overall width as on the existing standard cars, the already mentioned additional internal width or 'elbow space' of approximately five inches for each pair of cross seats.

Class E/1 tramcar with bodywork to design of experimental Class HR/1 tramcar 1852, with Alpax upper deck framing. (H Nicol/National Tramway Museum)

HR/1 upper saloon with upholstered seat backs. (© TfL from the London Transport Museum Collection)

HR/1 lower saloon seating. (© TfL from the London Transport Museum Collection)

"Special attention has been given to improving the means of ventilation. It is estimated that during hot weather it will be practicable [in the lower saloon] to supply approximately 300 per cent more fresh air per unit of time per passenger than with the existing arrangements. Separate provision is made to ensure that during inclement weather, when the opening of the main ventilators may be objected to by some passengers the air will be changed entirely without draught at least once every two minutes. For the upper deck a constant regular change of air is effected by extractors in the ceiling, the majority of which are arranged to work in conjunction with the saloon lamps. The large side window panels are supported from each side of the roof by a unique method, so that within the limits of maximum opening they may be adjusted to any position considered desirable to suit the weather. The supplementary ventilators above the staircases will be appreciated by passengers occupying the alcoves.

"Tests which have been carried out to ascertain the suitability and efficiency of the lighting arrangements used for the Pullman cars now in service have shown these to be practically ideal for London tramway traffic requirements, and consequently only minor modifications have been made on the new cars.

"Externally the special cars are painted and finished to match those of the existing Pullmans. Internally however, considerable modifications have been made, principally to accord with the modern tendency in the direction of simplicity in both form and applied colour decorated work. In the lower saloon the timber framing and panelling is of mahogany, with panels of light oak, whilst below the level of the seats the woodwork is of a dark rosewood colour. The side framing and cornices above the top edges of the windows together with the whole of the ceiling are finished to a plain surface with glossy white enamel.

"For the upper deck a generally similar scheme of finish is used, with the exception of the side framing and panels, which are finished in shades of ultramarine blue and antique oak respectively..." [70]

The outward appearance of these tramcars still bore a resemblance to the existing Class E/1 tramcars.

Meanwhile, a programme had been proposed at the LCC Highways Committee meeting of 29 January 1929 for the replacement of 50 tramcars; a further 304 were also needed.

The form of construction of the first of the experimental cars, No. 1852, was used for the tenders reported to the Highways Committee on 14 September 1928. Tendered for were 50 replacement bodies for the single deck subway tramcars which were to be rebuilt as double deck [71]. They were to have the characteristic wider central lower deck pillar for a steel channel section and the upper deck framing would be of metal sections. Although the tenders received were above the budgeted costs, the Tramways Department pressed for the lowest tender, from English Electric, to be accepted, reasoning that tooling and other costs had caused the discrepancy [72]. This order was for the construction of bodies for Class E/1 tramcars, fleet numbers 552 to 601.

The design and construction method of the body of the second experimental tramcar, 1853, with steel lower deck body and upper deck metal framing similar to that supplied for 1852, was the basis of the requests for tenders of 17 March 1929. These were received and published on 30 July 1929.

Orders were placed for the construction of 50 Class HR/2 and 100 Class E/3 tramcars [73]. The Class HR/2 tramcars were to have four motors, two in each bogie, mounted on each axle in equal size wheel bogies; the Class E/3 tramcars would have two motors, one in each bogie, mounted on the axle with

LCC experimental HR/2 tramcar 1853 with steel lower saloon body and Alpax upper deck framing. Note plough carrier attached to the truck. (© TfL from the London Transport Museum Collection)

larger diameter wheels, with an unmotored axle with smaller diameter wheels in the maximum traction type bogies.

Thus, in the time period between the request in November 1927 for the two experimental tramcars and July 1929: tenders were raised for parts for construction in April 1928; tenders were issued in September 1928 for 50 new tramcar bodies to the first new design, even before the two tramcars were completed and delivered in February 1929; and by July 1929, orders for 150 new tramcars to the second new design were in place.

In October 1929, within a few months of implementing these developments and orders, the LCC Tramways Department were requesting funding for another experimental tramcar:

"In view of the present day trend towards the improvement in design and appearance of public service vehicles... it appears desirable to improve upon the type of car recently approved by the Committee for hilly routes.

"Improvements at present in mind would include enclosed vestibule ends as an integral part of the structure, comfortable seating, adequate lighting, ventilation etc. and the general appearance of the exterior."

This new experimental tramcar was for consideration at the Highways Committee meeting on 24 October 1929; the Committee approved the proposal [74]. The Committee Minutes explained: "the proposal to construct an experimental car is based on the desirability of exploring possible new methods of construction and equipment." [75]

This experimental tramcar was to become LCC No 1.

More LCC tramcar orders

The LCCT's request to the Highways Committee for the 1931-1932 programme was for 150 tramcars intended for the Grove Park extension and to replace the remaining A and D Class tramcars [76]. The total number was to be made up of 120 E/3 and 30 HR/2 tramcars. In addition, 50 new E/3 tramcars were to be provided by Leyton Corporation.

A report of 5 February 1931 by the General Manager considered a revision of the 120 E/3 and 30 HR/2 numbers to meet the maximum service requirement of 1707 tramcars, by taking into account the number of modern type cars available – then 1,602. To this total could be added the 50 new Leyton Corporation cars. It was stated that 13 additional tramcars would be needed for the Grove Park to Eltham services, whereas using cars with larger seating capacity for Highgate Hill would save eight

LCC Class E/3 tramcar. (© TfL from the London Transport Museum Collection)

LCC Class HR/2 upper saloon. (© TfL from the London Transport Museum Collection)

LCC Class HR/2 lower saloon. (© TfL from the London Transport Museum Collection)

tramcars. The requirement for new tramcars was subsequently reduced to 60 Class HR/2 [77], thereby providing a total of 1712 modern type tramcars. In considering 30 HR/2 cars specifically for Highgate Hill services, a provisional sum was included to meet the cost of supplementary air brakes which the Ministry of Transport might consider desirable as an additional means of safety to motormen in handling these cars on the steep gradient.

Developments in tramcar bodywork construction technology

The move from timber construction to metal construction for tramcars was not a new development. After two serious underground railway fires in the early 1900s, the Board of Trade decreed that all future cars must be of metal construction. The same ruling applied to the Kingsway Subway. Dick, Kerr and Company and Brush & Co. built all-metal single deck tramcars in 1905; the cost at that time was about 20% more than comparable wooden tramcars [78].

Metropolitan Electric Tramways' experimental tramcar 318, 'Bluebell', which was completed early in 1927, was constructed with vertical pillars of spindle moulded ash encased in high tensile steel; most of the body panelling was in aluminium sheet. The roof was of laminated plywood which relied for strength on wooden cross members forming part of the ceiling. The main side members of the underframe were lightweight pressed steel Z sections, and the total weight was 12 tons. An additional aluminium domed roof was fitted by 1930 [79].

'Poppy', London United Tramways' experimental car 319, subsequently 350, entered service in April 1927 with a body of composite construction: metal panels on an ash body frame. It was constructed using standard components from the production line of NS type motor buses. Mounted on

Construction of experimental tramcar HR/1 No. 1852 with composite wood and steel lower saloon construction and Alpax upper deck framework. (© London Metropolitan Archives [City of London])

an underframe of Z section nickel steel, the tramcar weighed 16 tons and had a domed roof structure of aluminium [80].

LCC Tramways' experimental tramcar number 1852 of 1928 was constructed on a steel underframe with steel side truss plates, supplied by rolling stock manufacturer Hurst, Nelson and Company Ltd, and a lower deck timber frame and panelling. The central pillar was widened to allow for the installation of an inverted U-shaped continuous steel channel section from the side truss plates over the lower saloon roof. The top deck structure was of Alpax aluminium sections which supported the windows. Cladding

Construction of experimental tramcar 1853 with platform bearers and lower deck steel frame and panelling. (© TfL from the London Transport Museum Collection)

Construction of experimental tramcar 1853, interior of lower deck steel frame and panelling. (© TfL from the London Transport Museum collection)

Alpax upper deck framing used in the construction of experimental HR/2 No. 1853. (© London Metropolitan Archives [City of London])

was of sheet metal. The roof was a one-piece 3/8" thick heavy duty plywood board with material supplied by Tucker Armoured Plywood Co. Ltd, Crayford [81].

Hurst, Nelson's schedule of cars includes an order on 14 April 1928 for two car sets of swing bolster equal wheel bogies, roller bearings, one truck in each pair to take a plough, together with two underframes and one body frame [82]. These were fitted to 1852 and 1853.

The second experimental tramcar, 1853, differed from 1852 by having a steel underframe and lower deck body shell, in a steel frame box formation with straight sides [83]. The upper deck structure was of the same form as for 1852, made up of Alpax aluminium sections which supported the windows; cladding was of sheet metal, with the roof as a one piece, 3/8" thick, heavy duty plywood board material, also from Tucker Armoured Plywood Co. Ltd.

320, 330 and 331, the three experimental tramcars which were to follow 318 and 350 for the MET, were constructed with steel frames and steel and aluminium sheet panels [84].

The 100 production 'Feltham' or UCC tramcars were constructed with steel frames and steel panels to the bodywork.

In Birmingham, meanwhile, there were two experimental lightweight tramcars. The first, in conjunction with Short Brothers – aircraft builders who also at that time supplied bus and tramcar bodies - was from a design evolved in December 1928 to suit Birmingham requirements. It had a body of all-metal

 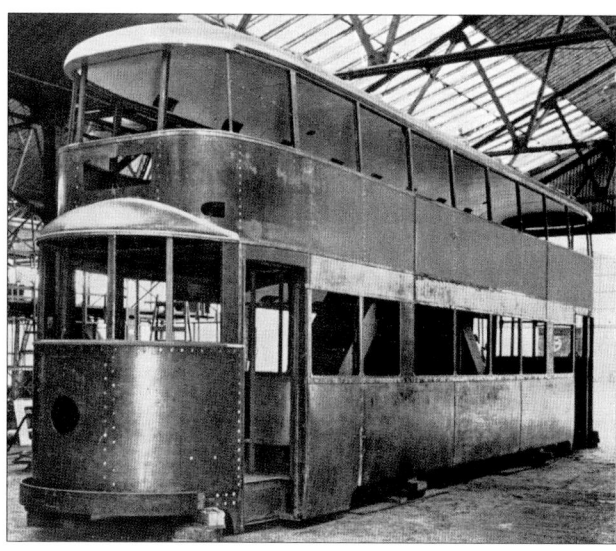

UCC Feltham-type tramcar body framing. (Tramway & Railway World)

UCC Feltham-type tramcar bodywork with steel panelling. (Tramway & Railway World)

Short Bros aluminium-framed Birmingham experimental tramcar 842. (Tramway & Railway World)

construction; aluminium alloys were used for the framing, roof and upper saloon floor, whilst the panels were of sheet aluminium. The underframe was constructed in steel, designed to take standard Birmingham bogies, brake gear and fenders. Wood was used to construct the lower saloon floor and platforms in order to reduce track noise. Internal lining panels below the waist rail were of aluminium faced with brown leathercloth. A body weight saving of two tons was achieved, giving a total weight, with lightweight trucks, of 13 tons 12 cwt. The body was delivered in October 1929 and entered into service in November 1929 [85].

The second tramcar was supplied by Brush. The body was completed in May 1930, with duralumin being used throughout. The underframe consisted of light steel members with duralumin flitch plates along each side forming the rocker panels; steel platform bearers were used. A feature of this car was its deeply domed roof. Its total weight was 12 tons 6 cwt, the body giving a 2 tons 10 cwt weight saving. It entered service in September 1930 [86].

Sheffield was also experimenting with weight reduction, with experimental tramcar number 370 being built at their Queens Road Works prior to June 1930. During 370's construction the airship R101 crashed at Beauvais. The aluminium body of the tramcar, then standing as a complete frame, had been built to the same stress formulae, and although the R101 enquiry was not completed, all the aluminium rivets in 370's frame were replaced by steel. On completion the tramcar was lighter by under two tons in comparison with the standard wood framed tramcars [87].

Birmingham 842 upper saloon seating. (Tramway & Railway World)

Birmingham 842 lower saloon seating. (Tramway & Railway World)

In early 1932, Edinburgh produced a lightweight experimental tramcar, 180, which was constructed with a steel underframe and duralumin framework, body stiffening and panelling [88].

Edinburgh experimental tramcar 180. (Tramway & Railway World)

Framework of experimental Edinburgh tramcar 180. (Tramway & Railway World)

Chapter 5
Experimental tramcar
Richard Sykes

Following approval for the construction given at the Highways Committee meeting of 24 October 1929, the London County Council Finance Committee agreed to funding of £5,000 for an experimental tramcar [1]. Within the tramways department the tramcar was to be described as the Experimental Tramcar, Blue Car or No. 1, but it was often heard to be referred to as 'Bluebird'.

Design

In the report produced by London County Council Tramways' General Manager, Mr Joshua Kidd Bruce, he said: "...improvements at present in mind would include enclosed vestibule ends as an integral part of the structure, comfortable seating, adequate lighting, ventilation etc (sic) and the general appearance of the exterior." Further explanation subsequently submitted was that "...the proposal to construct an experimental car is based on the desirability of exploring possible new methods of construction and equipment." [2]

Further design parameters were elaborated. "From charts of traffic density during the day it can be seen that a 65 seating capacity car with liberally designed standing accommodation would meet economically the traffic demands.

"This reduced seating capacity offers facilities to design a tramcar giving greater comfort and more in harmony with modern requirements.

"With the average seating capacity known and the maximum overall dimensions defined, the car takes the shape of a rectangular box about 36 feet by 7 feet 3 inches by 15 feet 3 inches high.

"Increasing the slack hour passengers is essential and it is the comfort of the modern motor coach that sets the standard of ease demanded by the passenger. Light weight, low floor, low maintenance costs, low energy consumption, high acceleration and retardation, and less noise are among the many improvements which are essential in new cars." [3]

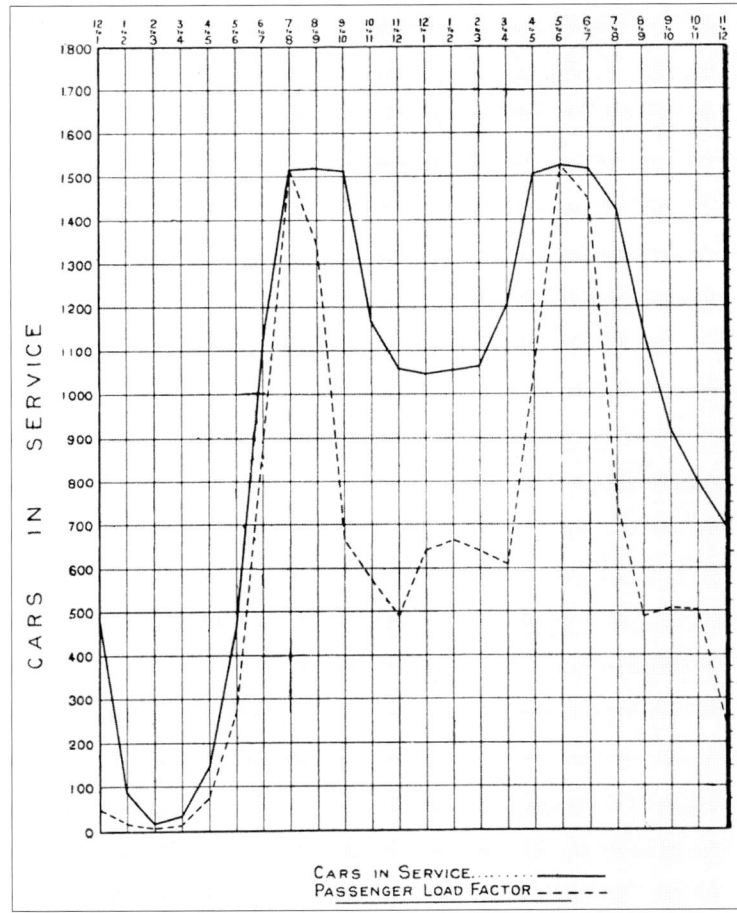

LCC tramways passengers carried during the day. (Tramway & Railway World)

Although this was not explicitly stated, the tramcar would need to be of metal construction to meet the safety requirements for use on routes through the Kingsway Subway.

In defining the dimensions, the length of the experimental tramcar was restricted to 36 feet due to the design of the LCC tramway depots. High land values and the use of the conduit system of current collection influenced the position and layout of the depots. To save space and eliminate pointwork, traversers were used which, along with the building, had been designed for tramcars up to 36 feet long. The accommodation of longer tramcars would have necessitated the rebuilding of traversers and at least the moving of some roof supports [4] [5].

LCC Stamford Hill depot. Traverser and roof supports limiting tramcar length.
(W A Camwell/National Tramway Museum)

A Manager's Report of October 1929 was entitled *As to an experimental type of double decked bogie tramcar for service on rails with wide track spaces.* The width of the tramcar was only to exceed that of the latest Class HR/2 and Class E/3 tramcars by half an inch, but with a length of 36 feet this tramcar was to be nearly 1½ feet longer. The overhang of the bodywork when turning corners would therefore be greater, which could entail the need for modification of trackwork for clearance between passing tramcars [6].

In order to decide on some of these design parameters, various possibilities were considered before a final approach was reached.

"What type of car would be most suitable to the metropolitan area will always be a debatable question. The single deck car either of the two unit articulated type or long single deck central entrance American type car have all the advantages of lightness and economical operation. However to meet the average loading of 65 seated passengers, the physical dimensions of single deck cars would prohibit their use, owing to the limitation of permissible overall length and width on the permanent way and at depots."

"The double deck car can be divided into three types - central entrance, front exit and rear entrance with controlled doors and the single exit and entrance. The central entrance double deck car loses a great many of its advantages owing to the very inadequate seating layout which can be provided in the lower saloon and the necessity of providing an extra conductor to control the entrance. The front and rear exit cars are good for certain traffic densities but where headways are maintained analogous to a conveyor system, the time lost in door operation would of necessity decrease the schedule speeds. Air operated doors have a time lag of 3½ seconds, and to this must be added the psychological effect on passengers who are inclined to remain seated until the doors open. For fare collection the front exit offers added facilities to the short distance passenger to avoid paying fares.

"The single entrance, with all its advocates for and against would seem to offer the best solution of the type of car for London's requirements." [7]

At the same time as work on the experimental tramcar was progressing, the final Report of the Royal Commission on Transport was produced by the Special Committee on Traffic. The general conclusions were not in favour of the retention of tramways; but the opposite view was held by the Highways Committee, who in February 1931, after discussing the Final Report, resolved to disagree with its

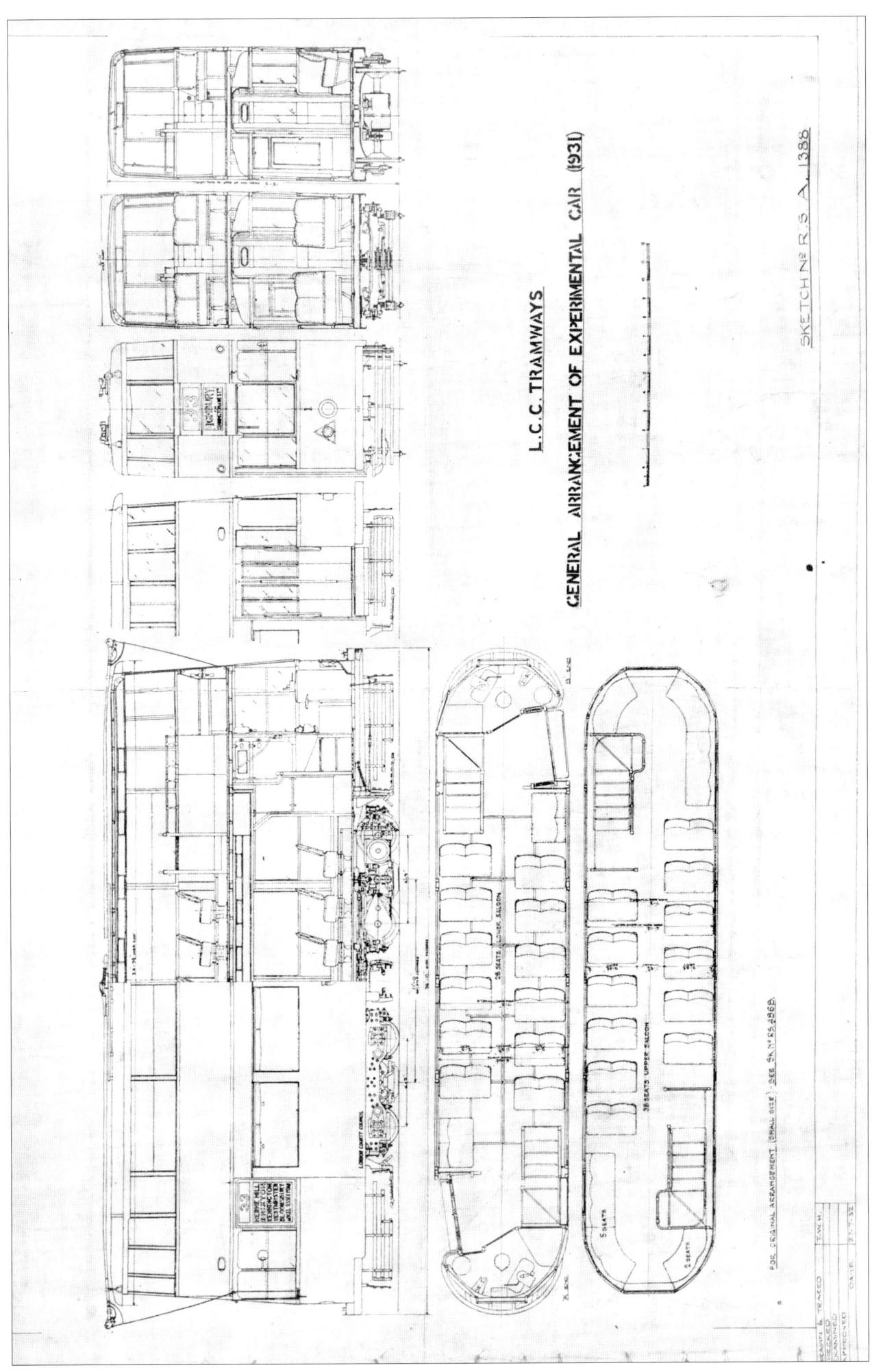

General arrangement drawing of the experimental tramcar. (National Tramway Museum)

recommendations. In the Committee's opinion it was impossible to foresee the time when, by reason of safety, fare standard, general comfort, hygiene and expedition, it would be practicable to replace the tramcar by any known alternative means of transport [8].

The bodywork was designed using the experience gained from the experimental tramcars 1852 and 1853. The lower deck frame and side panels were of steel as with 1853. The headstock design was however modified to eliminate the step from the platform to the saloon, to provide a flat floor the full length of the tramcar. Side steelwork framing was also extended to upper saloon window sill level. The upper deck windows of Alpax aluminium based framing and single piece composite plywood roof were to the principle used on 1852.

Construction

Up to the end of March 1931, capital expenditure on construction had been £641, which was the figure for the supply of two modified Class 6 trucks, with equal diameter wheels, and parts from EMB. The Class 6 chain-driven bogies, EMB order number 8964, were ordered in September 1930 and delivered in March 1931 [9]. For these two trucks £629 had been sanctioned for "two motors and associated equipment" from Metropolitan Vickers in December 1930 [10]. One motor in each truck was intended to drive both axles by chains.

Hurst, Nelson cast headstocks and steelwork on the experimental tramcar. (LCCT official photograph/LCCTT Collection)

In early September 1931, an order was placed with Hurst, Nelson and Company to supply steelwork - two body bolsters, two main side plates and sills and two headstocks cast from a pattern which was to become LCC property [11]. The total quoted value of this order was £59.30.

On 17 September, the Highways Committee agreed to postpone work on the experimental car as the Tramways Department, along with other LCC departments, as highlighted in a previous chapter, was required to reduce its budget for the year. The General Manager indicated in a report for the Highways Committee [12] that £641 had been spent in the financial year 1930/1 and £880 in the 1931/2 financial year, making a total of £1,521 up to the date of suspension of operations. According to the Manager's estimate, £3,000 was to be saved during the year ending March 1932 by halting the work on this tramcar [13].

At a Highways Committee meeting in December 1931 it was agreed that the construction work should recommence, the Finance Committee having endorsed future expenditure of up to £3,600 [14].

In the capital estimates for 1931/2, the figure for the construction of the experimental car by 31 March 1931 was £800, to be followed by another £3,000 by the same date in 1932 and a further £1,200 by 30 September 1932. The figures were revised for the 1932-3 capital estimate to £4,000 by 31 March 1932, and a further £1000 by 31 March 1933 [15].

Financial year	Budget total by 31/3/31		Budget total by 31/3/32	Budget total by 30/9/32	Budget total by 31/3/33
Budget estimate 1931/32	£800		£3,800	£5,000	
Revised Budget estimate 1932/33			£4,000		£5,000
	Actual total by 31/3/31	Actual total by 30/9/31			
1930/31	£641				
1931/32		£1,521			

The rate of progress with the experimental trucks and on construction of the bodywork can be gauged by the material purchases, some of which can be identified from records of quotations and are summarised in the table below [16].

Since the main frame steel work was only ordered just prior to the cessation of work, a notional start date for the bodywork is considered as week commencing November 9, 1931, when this steelwork could be expected to be in stock.

Minute or Quotation Date	Item Details or Quotation Approval Date(Q)	Year Week Number	Construction Progress Week numbers (from notional start 9/11/31)
September 1930	Modified Class 6 trucks ordered	36-40	
March 1931	Delivery of Modified Class 6 trucks	9-12	
4 September 1931	Headstocks, body bolsters, side plates (Q)	36	
17 September 1931	Suspension of work due to budget restrictions	38	
5 November 1931	Request for resumption of construction work	45	
21 November 1931	Large section structural plywood for the roof and upper deck floor (Q)	47	2
3 December 1931	Alpax metal upper deck window castings (Q)	49	4
7 December 1931	Alpax roof cowls Platform doors and steps (Q)	50	5 (year week 53 = week 8)
27 January 1932	Ceiling panels (Q)	4	12
February 1932*	Air brake equipment Bucket revolving double and single seats for both the upper and lower saloons Electric gong 80 square yards of plain blue linoleum (Q)	5-8	13-17
26 April 1932**	12 double transverse swing back seats for the lower saloon and 2 driver's adjustable seats (Q)	17	26
5 May 1932	To be available for presentation to the Highways Committee at Charlton	18	27

* In February 1932, the Highways Committee sanctioned the expenditure of £708 for this work [17]; it is noted that the electric gong set was ordered for one platform only.

** A quotation for endorsement by the Highways Committee dated 26 April 1932 for seating, which included the 12 lower saloon and 2 driver's seats, was presented to the Committee on May 2 1932 [18].

Following unsuccessful trials, the experimental trucks intended for the tramcar were not used and the available trucks intended for Class HR/2 tramcar No. 160 were redirected for use for LCC No. 1 [19]. The air brake cylinders, secured to the body, were connected to the existing brake linkages fitted on the trucks.

Four Metropolitan Vickers 35hp motors of type 109Z were fitted.

The controllers were Metropolitan Vickers OK 37B camshaft type with EMB interlock air valve box Type AJ located on the controller top. This EMB fully interlocking braking system for the air-wheel and air-track brakes was probably being applied in use for the first time with this tramcar [20].

All LCC bogie tramcars initially had castings attached to one of the trucks to carry the plough assembly, used for current collection from below the road surface. The top of the plough was fitted with two copper shoes which supplied the electricity for the tramcar when in contact with collector bars, installed

Lower deck steelwork and upper deck aluminium window framework on the experimental tramcar. (LCCT official photograph/LCCTT Collection)

Experimental tramcar upper deck aluminium framework and single sheet plywood roofing in front of the body with upper deck steel sheet applied. (LCCT official photograph/LCCTT Collection)

EMB Class 6A truck used on the experimental tramcar. Direct gearing of two motors, each mounted on a separate axle, on a truck diverted from fitting to the intended HR/2 No.160. (M C Crabtree)

Central Repair Depot, Charlton. Works staff with LCC No. 1. (LCCTT Collection)

above the channels of the plough carrier. These ploughs, described as 'sliding head contact ploughs', could move laterally across the plough carrier, which also enabled removal of the plough [21].

In the 1920s the LCC tramways continued to have problems with broken ploughs. Some failures were considered to be due to the hunting movement of the bogies which could throw the ploughs about the carriers. Others occurred when a tramcar had a flat spot on a pony wheel on the truck with the plough carrier; this could subject the plough to excessive lateral and vertical movement [22].

In 1929, Class E/1 tramcar No. 1565 had a plough carrier assembly fitted experimentally to the solebar. Trials with this body-mounted carrier proved successful. The subsequent production carrier consisted

Body mounted plough carrier of the type fitted to Class E and Class E/1 tramcars. (© London Metropolitan Archives [City of London])

Body mounted plough as fitted to the experimental tramcar. (National Tramway Museum)

of a metal framework containing the plough carrier channels and electrical bus bars as a self-contained unit which could be bolted to the base of the tramcar [23]. The new Class E/3 and Class HR/2 tramcars were constructed with body mounted plough carriers of a different design [24]. A body mounted plough carrier was modified to fit the wider underframe of the experimental tramcar.

LCC Tramways produced leaflets and posters to publicise tramway services generally and for attractions that could be travelled to by tramcar. Changing tramcar livery to red, in conjunction with their passenger related improvements, was used from 1926 to publicise the 'Pullmanisation' programme as a further promotion for the tramways.

It must have been decided that the experimental tramcar would have a different livery to the 'Pullman' cars. To evaluate styles and colours, a one inch to the foot scale model was constructed at the Central Repair Depot [25] [26]. It is suggested that the blue colour was initially based on the livery of 'Bluebell', MET's experimental tramcar No. 318. The model, considered to have been constructed in late 1931 or early 1932, can currently be viewed at the London Transport Museum, Covent Garden.

The final colours were described as Royal Blue with a groundwork of Ivory White, enriched at certain points by the addition of gold lining, with parts liable to traffic damage finished in black [27].

The model of the experimental tramcar was given the fleet number 1.

The completed tramcar was shown to the Highways Committee at the Central Repair Depot on Thursday 5 May 1932, which was also the press day.

The Metropolitan Stage Carriage number 6059 was fixed to the bulkhead above the platform. This number was issued on 10 June 1932, re-issued on 27 June 1933, and removed on 24 July 1934.

The tramcar weight was 20 tons 6 cwt which in comparison with the last batch of Class HR/2 tramcars at 17 tons 9 cwt would not enable the tramcar to be considered lightweight.

Associated experimental work

It was intended to design an equal wheel bogie car employing only two motors to drive the four axles, instead of the four motor design of the Class HR/2.

A pair of trucks was ordered from the Electro-Mechanical Brake Company (EMB), who had been supplying tramcar trucks as well as brakes since the 1920s. Described as Class 6 trucks modified to include radial arm drives [28], the trucks were ordered in September 1930 and delivered in March 1931. Both axles were driven by roller chains from one motor [29] [30]. In conjunction with this, the offer of Metropolitan Vickers to supply two sets of motors and control equipment suitable for a four axle drive experimental car was accepted in December 1930, as mentioned above [31].

Class E/3 tramcar No. 1986 was used for testing the trucks intended for the new tramcar. The underframe of No. 1986 was modified to take the trucks. The chains were unsatisfactory, as they kept breaking under load. A modified set of chains was fitted but these continued to give trouble, resulting in the decision to modify the truck by lengthening the radial arms of the axleboxes, which would bring the pivotal points of these nearer the axis of the motor armature. This was carried out in an attempt to

LCC model of the experimental tramcar with dark livery. Note the lettering in white. (© London Metropolitan Archives [City of London])

LCC model of the experimental tramcar with light livery. Note the lettering in black. (© London Metropolitan Archives [City of London])

LCC model of the experimental tramcar at the London Transport Museum. (© TfL from the London Transport Museum Collection)

LCC No. 1 at Charlton works. (LCCT official photograph/National Tramway Museum)

Experimental bogie intended for the experimental tramcar, with chain drive to the two axles from the single centrally mounted motor.
(© London Metropolitan Archives [City of London])

Modification to the underframe of LCC No. 1986 for trials with experimental bogies.
(© London Metropolitan Archives [City of London])

prevent the chains 'snatching' when the car was very lightly or heavily loaded, but the modification was also unsuccessful and the trouble persisted. It was then noticed that the car bolsters were becoming damaged, which in turn affected No. 1986's underframe, resulting in the car vibrating badly when rounding curves. It was at this point that the experimental trucks were discarded [32].

On a quotation presented to the Highways Committee in February 1932, relating specifically to the experimental tramcar, revolving seats are listed for both the upper and lower saloons. The supplier G D Peters promoted Luxury Coach seating for tramcars with their Auto Controlled Reversible Seat. Bucket revolving seats were already on trial on three Class HR/2 tramcars, arranged transversely with a double seat and a single seat either side of the aisle [33]. In the literature, dimensions for the double seat show that it was able to rotate within a seat pitch of a minimum of 32½ inches. The seat was available for several types of pedestal base [34].

The order indicates the seating arrangement: 26 in the lower saloon and 40 in the upper saloon. The seats were not subsequently used on this experimental tramcar, there being an order with G D Peters on 28 April 1932 for the supply of transverse swing back seats [35]. The actual spacing of the upper saloon seating on LCC No. 1 is at a pitch of 27½ inches and the seating arrangement is 28 in the lower saloon, 38 in the upper saloon.

The media reception

Press reviews commencing 6 May 1932 were headed "A Tramcar Revolution", "New Luxury Tramcars", and "The Blue Tram", with heaters, armchairs and diffused lighting featuring prominently in the descriptions [36]. *The Sunday Times* commented: "Those of us who have sold our cars to pay rates and taxes may find comfort in the fact that Everyman now has his Rolls–Royce, and that we all may be Everyman." [37]

Rotating double seats.
(G D Peters/Tramway & Railway World)

LCC No. 1 at Charlton. (© TfL from the London Transport Museum Collection)

LCC No. 1 upper saloon. (© TfL from the London Transport Museum collection)

LCC No. 1 lower saloon. (© TfL from the London Transport Museum Collection)

Further publicity through *The Pullman Review* highlighted the streamlining effect by stating that in addition to the livery, parts that project on the standard tramcars such as vestibules, indicator boxes, destination boards and headlamps were built into the body [38].

In May 1932, *The Tramway and Railway World* and *The Electric Railway, Bus and Tram Journal* both described the new tramcar with virtually word for word reports based on the information and photographs supplied by Rolling Stock Engineer, Mr G F Sinclair [39] [40]: "New ideas incorporated in the tramcar express to some degree the ideals of Mr T E Thomas [LCC Tramways' General Manager from 1930] of what a tramcar should represent in external appearance, internal luxury, silent running, high speed and disposition of passengers. He has been led to the conclusion that these days passengers must be attracted, and that human nature being what it is the greatest attraction is personal comfort." [41]

The full text is given in Appendix 1.

LCC No. 1's first introduction to the transport industry was to managers at the Tramways, Light Railways and Transport Association Conference which was held in London that year. Mr Thomas showed the new tramcar to the delegates during their visit to Charlton Repair Works on 19 May 1932.

It is recorded [42] that No. 1 was at Camberwell depot awaiting trials and during the conference a delegate rode with 'smooth running' between Charlton and the Embankment [43].

In a paper given shortly after the launch of LCC No. 1, Rolling Stock Engineer Mr G F Sinclair indicated that he would advocate composite timber and steel for bodywork construction [44]. He stated that the London requirement differed from other operators as all-metal tramcars were needed for use through the Kingsway Subway.

LCC No. 1's influence on tramcar design

In an article on LCC No. 1, Mr John Price wrote: "Seldom indeed had so much new thought, and so many new features, been in evidence in a single vehicle. Looking back one can recognise in No. 1 the first appearance of many features later adopted in other towns, though not until several years afterwards." [45]

Mr Price identified these features as: straight through floors on the lower deck [46]; straight staircases [47]; enclosed platforms, for the comfortable accommodation of standing passengers [48]; air operated doors and folding steps; enclosed driver's cab and elimination of bulkheads; increased number of lower deck transverse seats; elimination of draught screens on the upper deck; and electric bells and heaters. [49]

Mr Robinson, Liverpool's City Electrical Engineer, who in 1932 was in an Acting Managerial capacity for their tramways, was impressed by the design of LCC No. 1 and actioned design modifications into a batch of tramcars currently under construction at Liverpool Edge Lane works, on a Works Order for tramcars 770-799. In 1933, after the completion of the first 12 bodies already being progressed, Mr Robinson modified the order for tramcars to what was essentially the LCC No. 1 design [50]. For numbers 782 onwards, smooth white ceilings and concealed lighting, large side indicators and shallow domed roofs were introduced, and the streamline livery feature was also copied. With an additional Works Order for tramcars 800-817, these 36 Liverpool tramcars, fleet numbers 782 to 817, were constructed with a steel underframe but with timber framing to the bodywork. Known as 'Robinson' or 'Cabin cars', they had platform doors, and the upper saloon dispensed with vestibules, as on LCC No.1. The tramcars were constructed in Liverpool Edge Lane works and the first of the series entered service in November 1933; the average cost of construction was £2600.

The trucks, with their air-wheel and air-track brakes, were also similar to the EMB Class 6 trucks used with LCC No. 1, which formed the prototype for these subsequent Liverpool products. The Liverpool electrical equipment and controllers were different, but the same air brake control system with the EMB interlock valve box Type AJ on the controller was used [51].

Later batches of tramcars for Liverpool of similar bodywork outline, the 'Marks Cars', did not perpetuate all the features; they were produced without platform doors, while upper deck vestibules were reintroduced.

Two LCCT drawings dated April 1932, described as being for a 1933 model, adopted the body design of LCC No. 1 with alternative stairway and seating arrangements. The tramcar design was with maximum traction trucks with the size of the larger diameter wheels reduced from 31 inches to 27 inches for clearance beneath the bodywork. This format was not put into production.

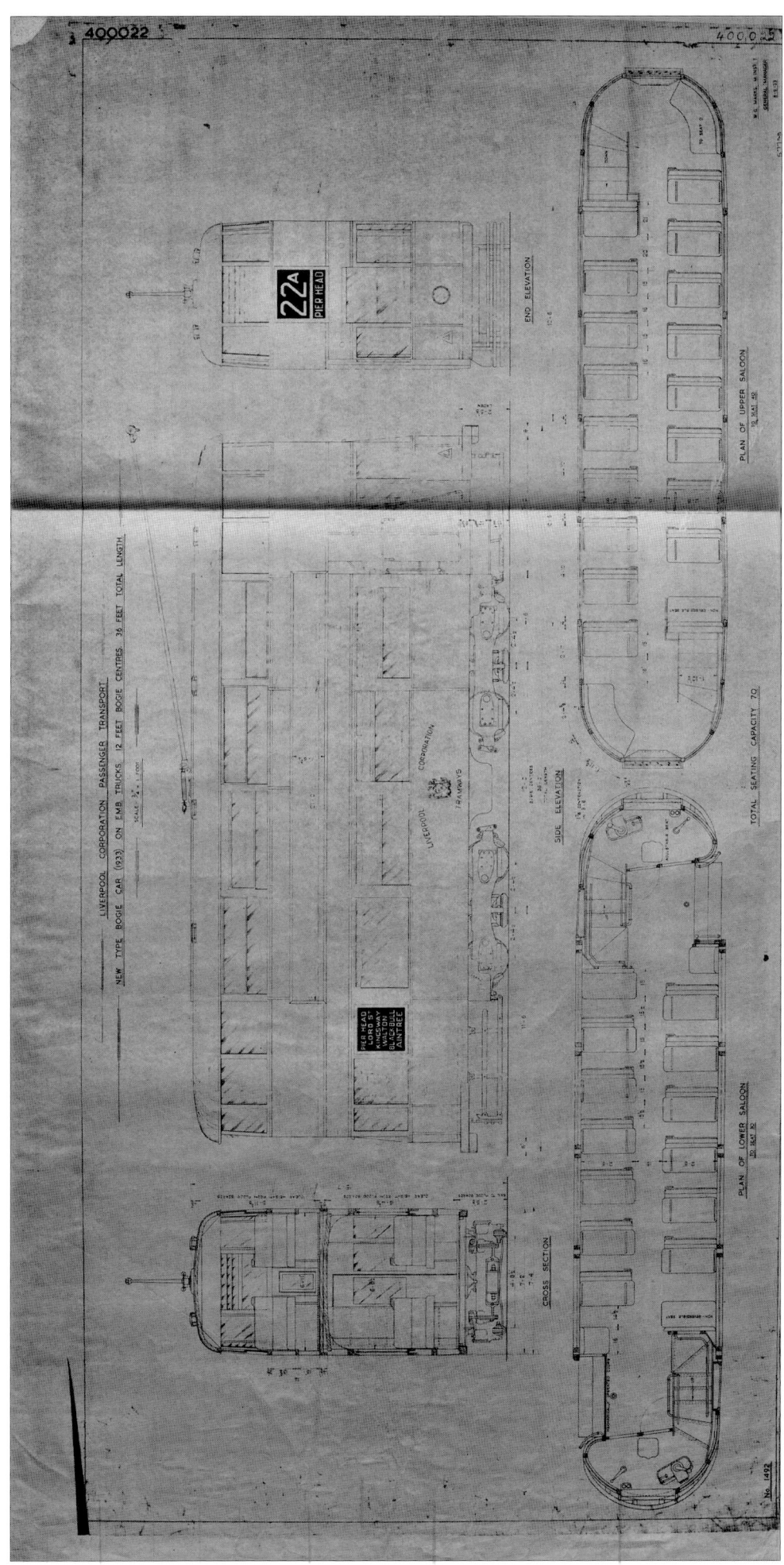

General arrangement drawing of Liverpool Robinson or Cabin class tramcar. (National Tramway Museum)

Robinson or Cabin class Liverpool tramcar in 1934, developed from LCCT drawings. (A D Packer)

Liverpool Cabin class tramcar upper saloon.
(M J O'Connor/National Tramway Museum)

Liverpool Cabin class tramcar lower saloon.
(M J O'Connor/National Tramway Museum)

A drawing produced under the auspices of the LPTB dated November 1933 was to make use of the Class 4 maximum traction truck as used on E/1 and E/3 tramcars. With a generally similar body design to LCC No. 1 this included revised stairways and greater seating capacity. The body was altered with a five window layout to the main saloons but losing the lower deck straight through floor in order to accommodate the larger diameter driving wheels [52].

This construction did not become part of the LPTB rehabilitation programme later in the 1930s, a programme applied to over 150 of the ex-LCC Class E/1 tramcars, where timber-framed modification of the upper deck construction was used.

Ultimately only the upper deck form of LCC No. 1 with Alpax window frames was used for replacement by the London Passenger Transport Board on a few ex-LCC Class E/1 and ME/3 tramcars [53].

After No. 1's introduction, subsequent bogie tramcar running gear designs used in the United Kingdom, with limited exceptions [54], featured equal wheel trucks.

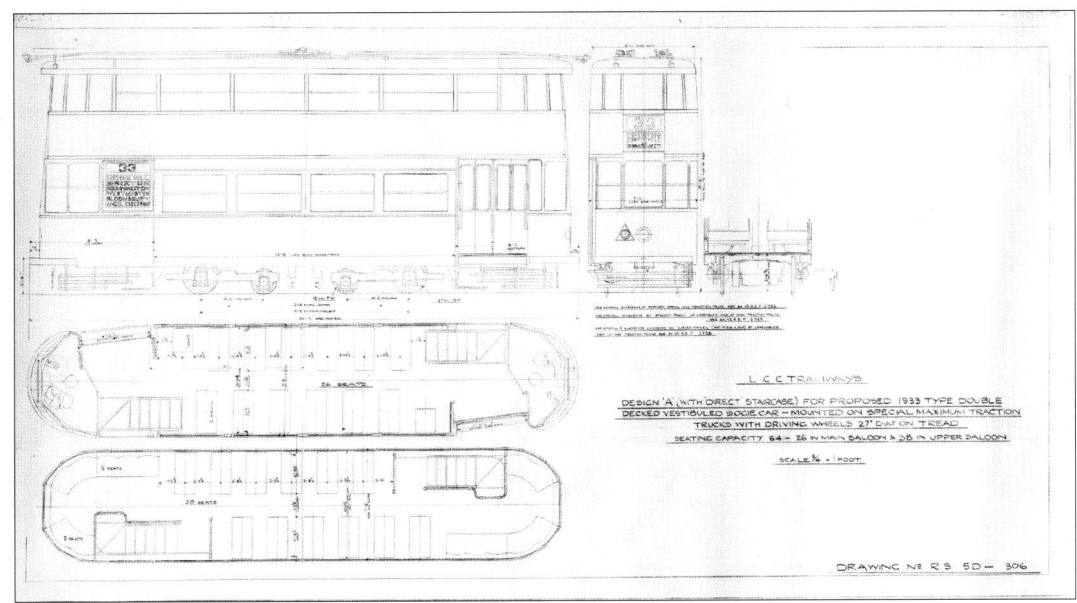
LCCT drawing 1932 with Maximum Traction trucks. (LCCTT Collection)

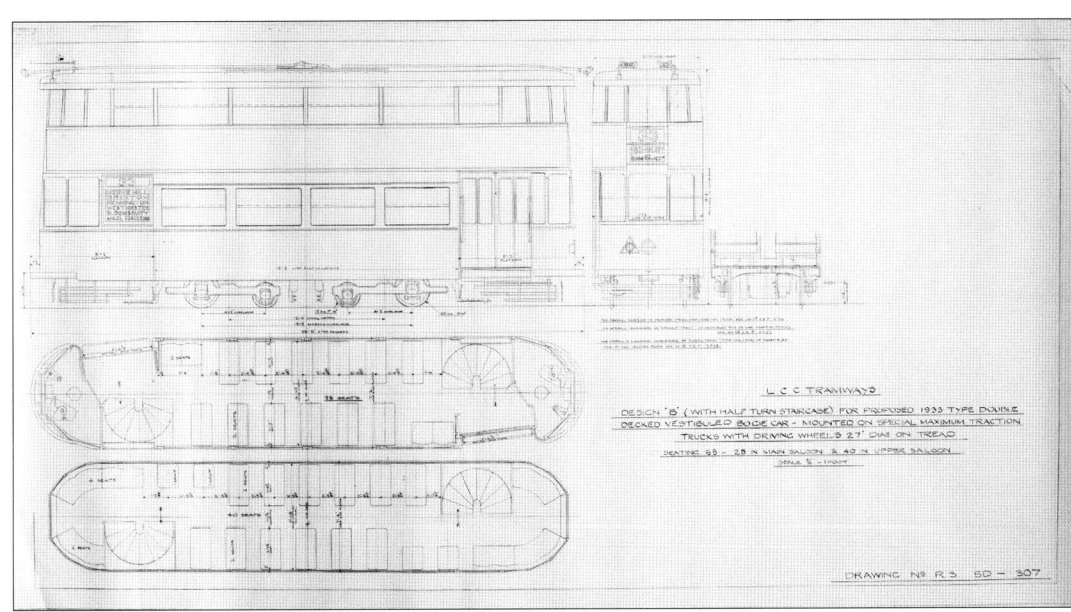
LCCT drawing 1932 with Maximum Traction trucks and modified staircase. (LCCTT Collection)

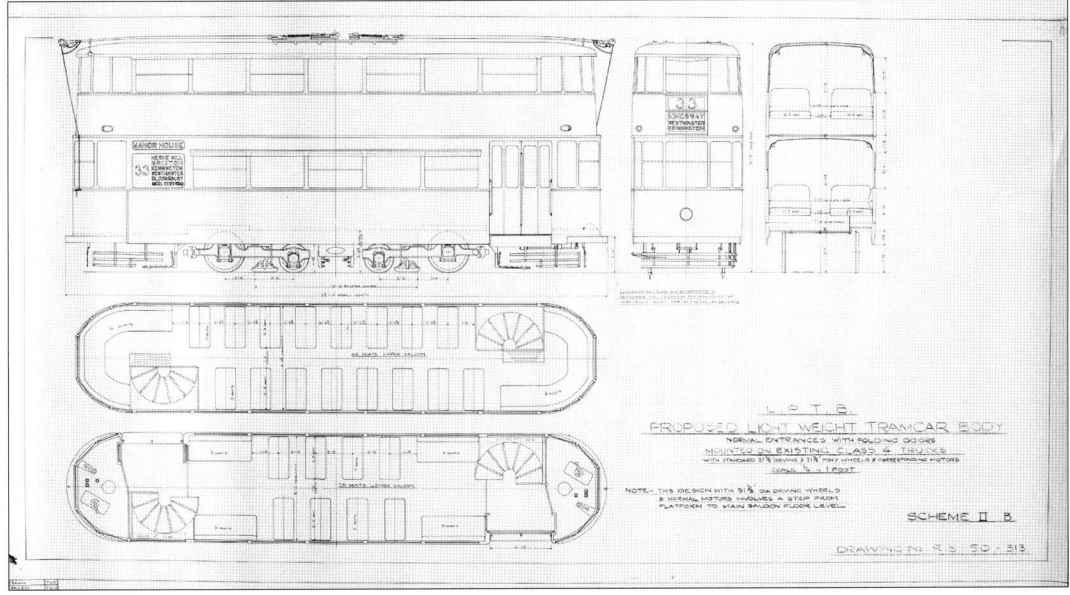
LPTB drawing 1933 with modified body and trucks. (LCCTT Collection)

Leeds Middleton Bogie tramcar 1933. (Tramway & Railway World)

Leeds Middleton Bogie tramcar upper saloon. (Tramway & Railway World)

Leeds Middleton Bogie tramcar lower saloon. (Tramway & Railway World)

It has been suggested that new tramcars in Bombay, manufactured by English Electric and delivered in 1932, owed design features to LCC No. 1. However, these tramcars could be considered to have been built on more traditional lines with timber – teak - for the body framework on steel underframes. The roof also differed from No. 1 in being of domed shape and of metal construction [55].

Bombay's LCC Class tramcars, Nos. 651-708, had actually been built in 1921-5 and were mounted on Brush maximum traction LCC-type bogies, and were reputedly designed after a study of the LCC Class E tramcars [56].

LCC No. 1 was not the only prototype to influence developments in new tramcar construction. Tramcar bodywork with central platforms for entrance and exit on the principles of another London prototype, Metropolitan Electric Tramways 331, was also introduced in the 1930s, with many features complementary to those on LCC No. 1, such as driver's cabs, platform doors and straight staircases. These types of tramcars were delivered to several tramway systems, including Blackpool, Darwen (Lancashire), South Shields and Sunderland.

Platform doors and separate driver's cabs, together with a lack of bulkheads and screens, could be found on other new tramcars, in addition to the Liverpool 'Robinsons' constructed after 1932, including those for Glasgow, Leeds and Liverpool. These tramcars also had traditional end rather than central platforms, along the lines of LCC No. 1. The Leeds Middleton Bogie tramcars had air operated doors in addition, with the door motor located as on LCC No. 1 [57].

Chapter 6

No. 1 in London

Lynn Wagstaff

On May 13, 1932, *The Electric Railway, Bus and Tram Journal* reported:

"NEW CAR TO BE INSPECTED BY CONFERENCE DELEGATES "

Members of the Tramways, Light Railways and Transport Association, who meet in Conference in London from May 18th to 20th, will have an opportunity of inspecting [the] new car when they visit the Charlton works on May 19th, and we have little doubt that after their inspection they will agree with our own opinion that tramways, far from being obsolescent as was suggested in the Final Report of the Royal Commission on Transport, are inherently capable of providing for their patrons greater travelling comfort and a larger measure of safety than any other form of street vehicle."

The Journal further explained: "… [the members] will be received by Mr T. E. Thomas, the General Manager, who will open a short discussion on the new tramcar and what is involved in its design and equipment." [1]

A week later the same publication reported on the events of the Conference, including the aforementioned visit: "Yesterday afternoon the Conference delegates visited the Central Repair Depot of the London County Council Tramways Department. Tramcars of the L.C.C.'s most modern type conveyed the party to and from Charlton, the new experimental car being brought into service for this purpose. Members were loud in their praise of the luxurious comfort and smooth running of the modern rolling stock." [2]

A correspondent for *Omnibus Magazine* also reported on the smooth running between the Embankment and Charlton [3].

On the last day of the Conference the delegates visited Fulwell Depot and Works, where Mr C J Spencer showed them the new LUT trolleybuses. At the end of the visit, Mr Spencer took them to Gunnersbury Station by Feltham tramcar [4].

During construction, reports and requests by the Tramways Department, and the Highways Committee Minutes, all referred only to 'the experimental tramcar'. Even when the work was completed, as the above reports show, it was still simply 'the new tramcar' or 'the new experimental car'. How did the LCC's newest tramcar come to be allocated the fleet number '1'?

No. 1 at Charlton in May 1932. Mr E H Edwards and Mr A de Turckheim, Chairman and Secretary of the Tramways, Light Railways and Transport Association, with Mr T E Thomas, the LCC Tramways General Manager. (Tramway & Railway World)

Travel pass for delegates to the 1932 Annual Conference, including travel on LCC No. 1. (LCCTT Collection)

The two experimental tramcars completed in 1929 were allocated fleet numbers 1852 and 1853, following in sequence from the previous Class E/1 delivery, which ended with fleet number 1851. The next production batch of 150 tramcars commenced in sequence at number 1854 [5].

After reaching fleet number 2003, the numbering of new LCC tramcars recommenced at 101 when in 1929 the Class HR/2 tramcars took up this sequence of numbers, which had been vacated by Class B tramcars delicensed since 1925. By the time new Class E/3 tramcars were delivered in 1931, more numbers in the same sequence had been vacated by Class C tramcars delicensed by October 1930; thus the Class E/3 tramcars were able to continue the numbering up to 210.

The new experimental tramcar could have been expected to follow in sequence after the Class E/3 tramcars, and be number 211. However, the last of the Class A tramcars of 1903, fleet numbers 1 to

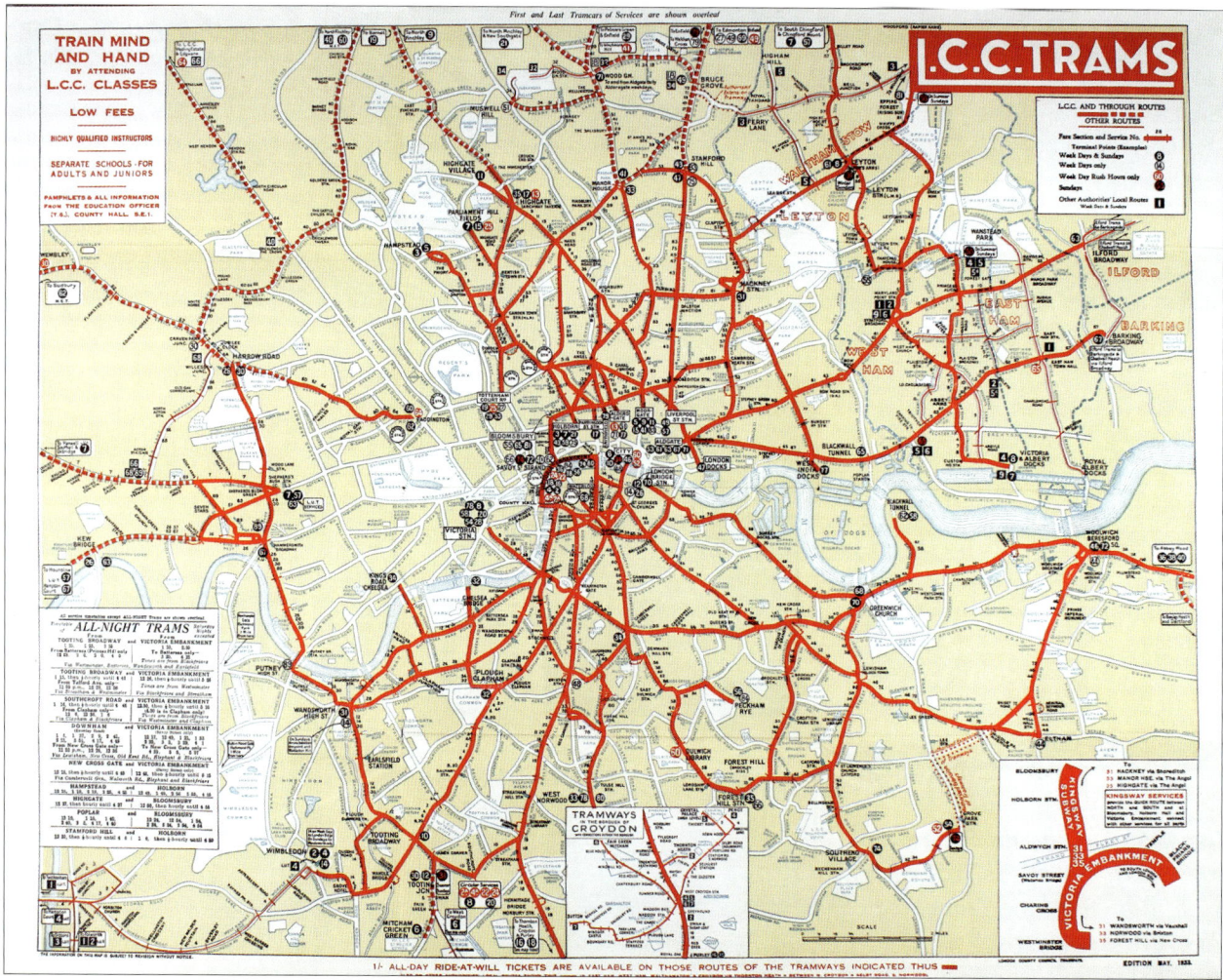

LCC Tramways map of the system when No. 1 entered service. (©TfL from the London Transport Museum Collection)

100, were delicenced in October 1931 [6], leaving the number 1 available. It is not known who took the decision to give this number to the new tramcar.

At Camberwell Depot awaiting trials [7], and following night time trials after its launch in May, LCC No. 1 was allocated to Holloway Depot and went into service on 10 June 1932. It worked several south side routes on trial before settling down on the Kingsway Subway services 33 and 35 [8] [9], working routes 35 and 35A on weekdays and route 33 at weekends [10].

The routes and destinations of the subway services were as follows [11]:

Service 33

From the south, starting at WEST NORWOOD travelling north - Thurlow Arms - Tulse Hill Station- Herne Hill – Effra Road - Brixton – Kennington - Lambeth North - Westminster Bridge - Victoria Embankment - KINGSWAY SUBWAY – Bloomsbury - Theobalds Road - Rosebery Avenue – Angel - Essex Road - Newington Green - Green Lanes to MANOR HOUSE. Both conduit and overhead supply were used on this route.

Service 35

From the south, starting at FOREST HILL travelling north – Brockley Rise - Brockley Cross – New Cross Peckham - Camberwell Green - Elephant & Castle - Lambeth North - Westminster Bridge - Victoria Embankment - KINGSWAY SUBWAY – Bloomsbury - Theobalds Road - Rosebery Avenue – Angel - Upper Street – Highbury Corner - Holloway Road to HIGHGATE ARCHWAY STATION. This route was powered by conduit supply only.

Service 35A

From the south, starting at ELEPHANT & CASTLE travelling north - Westminster Bridge - Victoria Embankment - KINGSWAY SUBWAY – Bloomsbury - Angel Islington to HIGHBURY. This route number was to be withdrawn by 1938, the service continuing to operate as part of Service 35.

LCCT ticket recovered from No. 1. It was issued for services 13, 33 or 35. (National Tramway Museum)

On Service 33, LCC No. 1 was one of around twenty tramcars providing the basic eight minute interval service. The full journey from West Norwood to Manor House covered 11½ miles, and the running time was just under an hour and ten minutes. Around thirty tramcars provided the service on Service 35.

Timetable for Service 35, within which No. 1 operated on a regular weekday schedule. (National Tramway Museum Collection)

The Holloway crews both had their rest days on Wednesday, which therefore became No. 1's maintenance day, although it was often out of service at other times due to a lack of spare parts [18].

Evidence from the tramcar suggests that the aluminium framing of the upper deck end windows fractured quite soon after entering service. The windows were reinforced with a steel framing supporting, but not replacing, the defective material. Changes were also made to the service number box; it was widened to accept three characters, enabling 35A to be displayed. The fixed glazing above the destination equipment was also altered to a half drop opening window arrangement.

LCC No. 1 may have presented maintenance problems extra to those of the other tramcars in the LCC Tramways fleet. Its service brake was not the track brake but the air brake system, which operated brake blocks onto the wheels. These brake blocks will therefore have required more frequent replacement than those on HR/2 tramcars. Examination of the trucks has indicated that the brake gear and support beams required dismantling before the brake blocks could be replaced; this was probably a more lengthy operation than track brake shoe replacement.

LONDON COUNTY COUNCIL TRAMWAYS

General Order No. 709

NEW EXPERIMENTAL (BLUE) CAR No. 1

Motormen operating this car have been instructed to bring it to a standstill when men are working on the line, and employees are hereby instructed that they must not stand between the tracks on routes over which this type of car operates.

The car will operate on Service **33** via Manor House, Angel, Holborn, Kennington Gate, Brixton, Herne Hill and Norwood.

Also on Service No. **35** via Highgate, Highbury, Angel, Holborn, Elephant and Castle, Camberwell Green, New Cross, Brockley and Forest Hill.

23, Belvedere Rd., S.E.1   T. E. THOMAS,
July, 1932.   2227   General Manager

Warning notice to staff working on the tramway. (LCCTT Collection)

In the period 1932 to 1933, hand-painted advertisements were applied to the tramcar's upper side panels. 'Tramcars provide Comfort for Reading and Smoking' appeared on one side, with 'Go-as-you-please fares... 1/- All Day' on the other [19]. Using posters to publicise the benefits of travelling by tram was a recognised tactic, and here No. 1 became a mobile advertisement, drawing the eye with its promotional messages as well as its carefully-chosen shade of blue.

Along with all other LCC Tramways vehicles, ownership of No. 1 passed to the London Passenger Transport Board (LPTB), operating under the 'brand name' London Transport, on 1 July 1933 [20]. The London Passenger Transport Act of that year stipulated that the Board must "... so... exercise their powers... as to secure the provision of an adequate and properly co-ordinated system of passenger transport for the London Passenger Transport area." [21] For the time being, the former LCC tramcars continued to run along their accustomed routes.

LPTB No. 1, as it now became, is recorded as being overhauled at Charlton Works, the LCC's Central Repair Depot, on 28 March 1935; it was repainted, retaining the same blue and ivory livery but with London Transport gold fleet names and standard L T Johnston gold fleet numbers. It also carried advertisements for 'All-Night Tram Services'. The tramcar returned to service on 20 May, only to be sent for accident repairs nine days later; we do not know the nature of the accident. The repairs were completed on 18 June and No. 1 again returned to traffic [22].

A schedule for May 1937 shows that LPTB No. 1 was provided with a specific duty, 2624 from Holloway Depot, being identified in the schedule as Blue Tram Operation on Route 35 EX during the week and Route 33 EX at the weekends.

Another overhaul at Charlton Works is recorded as commencing on 22 October 1937. This time the work included repainting, and when No. 1 returned to traffic on 29 November it was in standard London Transport red and cream livery [23].

During the same year Mr Jackson, one of the car's regular drivers, died [24], after which it is claimed that No. 1 was rarely used [25].

Under London Transport (LPTB) ownership, and following the replacement of the Shepherd's Bush to Uxbridge tramway with trolleybuses on 15 November 1936, the Union Construction Company (UCC)-built Feltham tramcars were transferred to Telford Avenue, Streatham during 1937 and 1938 - the LUT

No. 1 showing Comfort advertisement. (LCCT official photograph/National Tramway Museum)

No. 1 showing 1/- All Day advertisement. (LCCT official photograph/National Tramway Museum)

No. 1 showing All-Night Trams advertisement. (TLRS Collection)

(26 & 27.5.37) 48

Day.	Route No.	Time-Schedule No.	Route.	Trams per hour. Peak am	Trams per hour. Peak pm	Trams per hour. Normal M.day	Trams per hour. Normal Eve	Duty Schedule No.	Allocation. Depot.	No. of Trams. am	No. of Trams. pm	Total No. of Trams. am	Total No. of Trams. pm	Remarks.
M—F.	33	785	West Norwood & Manor House Station	12	12	10	7½	2455 2620	Holloway Norwood	14 12	15 13	26	28	
M—F.	33 Ex	682	West Norwood & Victoria Embankment (Savoy St.)	2 trams	—	—	—	2620 ⎱ 2621 ⎰	Norwood	2	—	2	—	
Sat.	33	787	West Norwood & Manor House Station	10	10	10	10	2618 2632	Holloway Norwood	12 11	12 11	23	23	
Sat.	33 Ex	683	West Norwood & Victoria Embankment (Savoy St.)	2 trams	—	—	—	2632 ⎱ 2633 ⎰	Norwood	2	—	2	—	
Sat.	33 Ex	1669	West Norwood & Manor House Station	Journeys				2694	Holloway	1	1	1	1	Blue Tram Operation
Sun.	33	3923	West Norwood & Manor House Station	—	—	6	7½	2624 2631	Holloway Norwood	7 6	10 7	13	17	
Sun.	33 Ex	1669	West Norwood & Manor House Station	Journeys				2694	Holloway	1	1	1	1	Blue Tram Operation
M—F.	34	664	Chelsea (Kings Road) & Blackfriars	15	15	7½	7½	2249	Camberwell	13	14			
	34		Chelsea (Kings Road) & Camberwell Green	—	—	7½	7½	2611	Clapham	17	17	30	31	
Sat.	34	513	Chelsea (Kings Road) & Blackfriars	12	7½	12	7½	2592	Camberwell	11	12			
	34		Chelsea (Kings Road) & Camberwell Green	—	7½	—	7½	2605	Clapham	13	14	24	26	
Sun.	34	55	Chelsea (Kings Road) & Blackfriars	—	—	10	7½	2593	Camberwell	10	11			
	34		Chelsea (Kings Road) & Camberwell Green	—	—	—	7½	2603	Clapham	8	12	18	23	
M—F.	35	143	Highgate (Archway Tavern) & Forest Hill	10	10	7½	7½	2596	Camberwell	16	16			
	35		Highbury & Elephant & Castle	—	—	7½	—	2455	Holloway	14	14	30	30	
M—F.	35 Ex	1669	Highgate (Archway Tavern) & Forest Hill	Journeys				2694	Holloway	1	1	1	1	Blue Tram Operation
Sat.	35	4047	Highgate (Archway Tavern) & Forest Hill	10	10	10	10	2639 2618	Camberwell Holloway	16 14	17 14	30	31	
Sun.	35	4054	Highgate (Archway Tavern) & Forest Hill	—	—	6	7½	2636 2624	Camberwell Holloway	8 8	11 9	16	20	
M—F.	36	1072	Abbey Wood & Victoria Embankment (via Blackfriars Bridge)	13½	13½	10	7½	2630	Abbey Wood	28	28			
	38		Abbey Wood & Victoria Embankment (via Westminster Bridge)	13½	13½	10	7½	2646	New Cross	38	40	66	68	
M—F.	36/38 Ex	1657	Woolwich (Beresford Square) & Greenwich (South Street)	3 trams	—	—	—	2630	Abbey Wood	2	2			
			Woolwich (Beresford Square) & Greenwich Church	—	2 trams	—	—	2646	New Cross	1	—	3	2	
Sat.	36	1069	Abbey Wood & Victoria Embankment (via Blackfriars Bridge)	13½	13½	20	20	2271	Abbey Wood	25	25			
	38		Abbey Wood & Victoria Embankment (via Westminster Bridge)	13½	13½	—	—	2647	New Cross	43 & 2 Ex	43	70	68	
	36 Ex		Wickham Lane & Victoria Embankment (via Blackfriars Bridge)	1 tram	—	—	—							
	38 Ex		Wickham Lane & Victoria Embankment (via Westminster Bridge)	1 tram	—	—	—							
Sun.	36	791	Abbey Wood & Victoria Embankment (via Blackfriars Bridge)	—	—	7½	10	2648	Abbey Wood	8	18			
	38		Abbey Wood & Victoria Embankment (via Westminster Bridge)	—	—	7½	10	2656	New Cross	27	28	35	46	

Holloway Depot scheduling indicating regular schedules specifically for No. 1. (LCCTT Collection)

No. 1 at Holloway Depot, April 1938. (W A Camwell)

Internal memo associated with the movement of No.1 from Holloway Depot to Streatham. (LCCTT Collection)

No. 1 operating on Service 18 EX during wartime conditions. (TLRS Collection)

Service 16 – No. 1 at Streatham in 1948. (A D Packer)

No. 1 operating on Streatham Extra Service. (Masterman/TLRS Collection)

Felthams first, and the MET Felthams afterwards [26]. These tramcars also had air brakes, but with different trucks and a less complex arrangement for brake block replacement.

Trolleybuses had also begun providing replacements to tramway services in northern London and some were displacing trams from Holloway Depot [27]. LPTB No. 1 was declared surplus to requirements at Holloway and followed the transferred Felthams to Telford Avenue in April 1938 [28]. Again it is interesting that No. 1 is referred to on the transfer document as the 'blue car' – several months after it had been repainted in red. In addition, on its transfer to Streatham, destination blinds for the tramcar were fitted and marked: 'Streatham Blue Car no 1 LPTB June 9 1938 Front and rear.'

Operating from Telford Avenue, No. 1 was mainly used for rush hour 'extras' on route 16 EX, and, less frequently, on some or all of routes 10, 16/18 and 22. Driver Collins – London Transport appear to have used the title 'Driver' rather than the LCC's probable preference, 'Motorman' - remembered: "… It wasn't on service all day, only a peak-hour car, Norbury Extra's (sic) usually. They did try to put it into all-day service and several drivers had a go at it, but they were holding the service up…" He also remembered the passengers' preference for the tramcar, recalling: "… when I had it on Norbury Extra's (sic) they used to wait for it on the Embankment – and if you let it out it was really fast. For some reason they would not allow standing passengers on it, so the Conductors liked it too…" [29]

The 16 EX services were the ones described as Norbury extras. Tramcars travelled northbound from THORNTON HEATH POND or NORBURY – Streatham – Brixton - Kennington Road - Westminster Bridge - Victoria Embankment – returning southbound – Blackfriars Bridge - Elephant & Castle – WIMBLEDON. The 18 EX operated in the opposite direction, going north from WIMBLEDON [30].

LPTB ticket recovered from No. 1. Issued for services 16, 18 and 22. (National Tramway Museum)

The other routes were as follows [31]:

Route 10
TOOTING BROADWAY – Southcroft Road – Streatham – Brixton – Kennington – Elephant – SOUTHWARK BRIDGE. Some weekend services ran from Tooting Broadway to St George's Church only. The full journey took 48 minutes and services ran every four to 10 minutes.

Route 16
PURLEY – Croydon – Norbury – Streatham – Brixton – Kennington – Westminster Bridge – Blackfriars Bridge (VICTORIA EMBANKMENT). A 71-minute journey with services running every three to five minutes.

Route 18
PURLEY – Croydon – Norbury – Streatham – Brixton – Kennington – Elephant – Blackfriars Bridge (VICTORIA EMBANKMENT). Times were the same as route 16.

Route 22
TOOTING BROADWAY – Southcroft Road – Streatham – Brixton – Kennington – Westminster Bridge – Blackfriars Bridge (VICTORIA EMBANKMENT). The journey took 47 minutes and trams ran every four minutes.

In addition to service duties, No. 1 was used for a number of private tours. On 15 May 1938 it was hired by a party of enthusiasts for the first time; the hirers were the recently-formed Light Railway Transport League (LRTL). The tramcar travelled from its depot at Telford Avenue to Waltham Cross in time to start the tour at 2.15 pm. The League members enjoyed a 60-mile ride via Edmonton, the Kingsway Subway, Victoria Embankment, Brixton, Streatham, Norbury, Croydon, and the Purley terminus. Tea was taken there, after which the tram set off on its return journey to Waltham Cross. At Brixton the members were treated to a detour through the new one-way system at Vauxhall, which had been opened only that morning. No. 1's total distance for the day, including its 'commute' from and to Telford Avenue, was 105 miles [32].

Records indicate that the car was withdrawn from service at the outbreak of war in 1939 and stored at the back of Telford Avenue Depot, where it was surrounded by sand bags and used as an air raid shelter. By the beginning of 1942, bomb damage elsewhere had caused a shortage of cars, and on 31 March No. 1 was sent to Charlton Works for overhaul. It was reinstated on 18 May, and used on regular peak hour service on routes 16 EX and 18 EX to Streatham and Norbury [33].

Returned to storage pending an overhaul in 1944/5, by 1946 it had returned to use on peak hour services from Telford Avenue [34]

Ownership changed again in 1948, from the London Passenger Transport Board to the London Transport Executive (LTE). The tramcar was now LTE No. 1.

Another private tour is recorded as taking place on 3 July 1949, when The Southern Counties Touring Society hired the car for a trip to Lewisham, Grove Park, Eltham and Blackwall Tunnel [35]. By this time, the London tram routes were being taken over by motor buses, while redundant tramcars were being scrapped; and Leeds City Transport had expressed an interest in acquiring some of these surplus tramcars as part of their efforts to improve their fleet.

In December 1949 one of the surplus London UCC Feltham tramcars was loaned to Leeds, where trials were conducted. As a result, the Leeds Transport Committee agreed to buy what remained of the original 100 UCC Felthams, 92 tramcars in all [36].

Modern Tramway article reviewing the 1938 tour. (Copyright LRTA)

1938 LRTL tour at Waltham Cross. (W A Camwell/National Tramway Museum)

1938 LRTL tour at Red Deer, South Croydon. (TLRS Collection)

1951 LRTL Tour programme with No. 1 and a Feltham tramcar. (LCCTT Collection)

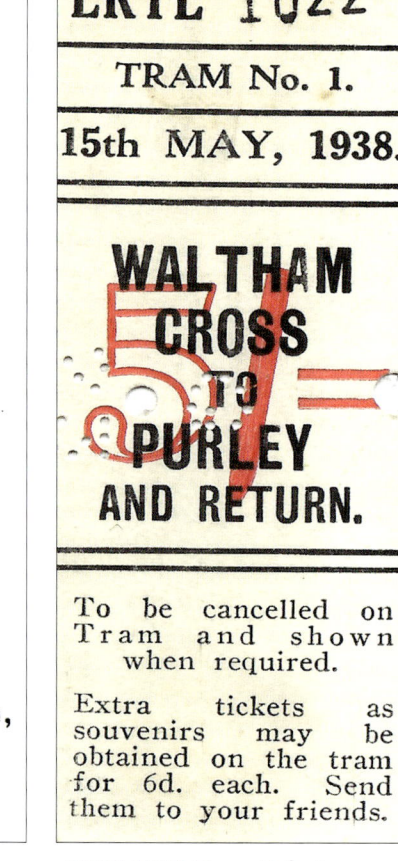

1938 LRTL Tour ticket. (LCCTT Collection)

SCHEDULE II
PROVISIONAL RATE OF DELIVERY

Tramcars Described in Part I of Schedule I	Oct.-Dec. 1950 Stage 1	Jan.-Mar. 1951 Stage 2	April-June 1951 Stage 3	July-Sept. 1951 Stage 4	Oct.-Dec. 1951 Stage 5	Jan.-Mar. 1952 Stage 6	April-June 1952 Stage 7	July-Sept 1952 Stage 8	Oct.-Dec. 1952 Stage 9	Total Cars
E1 (Pre.1920 Class)	56	92	64	13	56	3				284
E1 (1920 Class)	18					62				80
E1 (Croydon)						19	4			23
E1 (Walthamstow)						18				18
E1.(Sub)				21			25			46
E1 (E.Ham)							20			20
E1 (W.Ham)							31			31
E1 (M)							2			2
Special								1		1
Special purpose cars	3x	2x	2x	2x	2x	2x	2x	5∮ 1x	3∮ 1x	25
Total (A) cars	77	94	66	36	58	104	84	7	4	530
Tramcars described in Part 2 of Schedule I										
HR (No Trolley Gear)				51						51
HR (With Trolley Gear)								39		39
E3								76	67	143
(B) Total cars				51				115	67	233

x 17 Snow brooms
∮ 5 Stores cars
   1 Sand car
   2 Rail grinder cars

Schedule of tramcar disposals between October 1950 and December 1952. (LCCTT Collection)

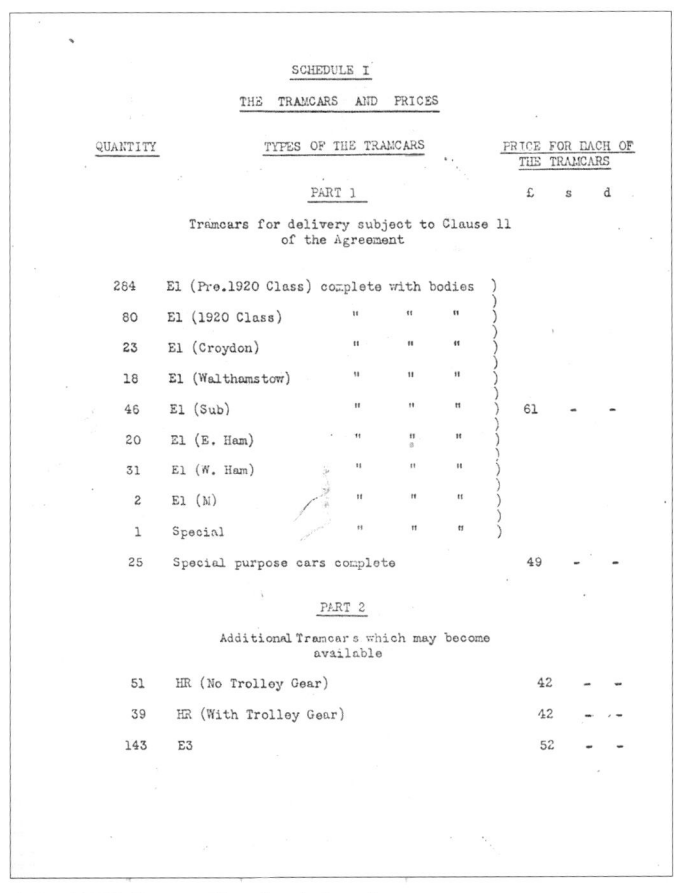

July 1950 Stage III schedule of tramcar moves. (LCCTT collection)

Meanwhile, an undated London Transport Schedule [37] had listed several hundred ex-LCC tramcars for disposal, including LTE No. 1, which along with most of the other cars was given a scrap value of £61. This Schedule indicated that the disposal was planned to take place in nine stages, the first in October-December 1950, and the ninth two years later. No. 1 was included at Stage 8, July-September 1952, implying that the tramcar could have continued in service in London until then.

However, the London Transport Executive's changed plans are revealed in another Schedule, dated 14 July 1950. This divides the withdrawal and transfer to Leeds of the 92 Felthams into three stages, beginning in October 1950 and ending in April 1951; it also documents the provisional inclusion of Tram No. 1 at Stage 3, but further states that No. 1 was, in the meantime, to be transferred from Brixton Hill Depot to Penhall Road, Charlton [38]. Only a mile away from the Central Repair Depot in Feltram Way, Penhall Road was the site where withdrawn tramcars were sent to be scrapped [39].

No. 1 was, thankfully, not destined to end its existence at Penhall Road. On 18 November 1950 a fire at Brixton Hill Depot destroyed the bodies of two of the UCC Felthams scheduled for transfer to Leeds. At a London Transport Supplies Meeting on 24 May 1951, it was stated that No. 1 was available for disposal and that Leeds City Transport wished to buy it. The tramcar's provisional inclusion on the list of tramcars to be transferred became a permanent one, and the sale was authorised [40]. Leeds paid £500 for the erstwhile 'Blue Tram' [41].

7 April 1951 was LTE No. 1's last day in service in London; driven by Stan Collins, it was used on an LRTL tour from Highgate to Purley [42]. After the tour, the tramcar was driven to Charlton in preparation for disposal to Leeds; whilst stored at Penhall Road, its plough, trolley gear and trolley hooks were removed [43].

No. 1 was then transported from Penhall Road to Leeds, arriving there on 19 June 1951.

The last word in this chapter must go to Stan Collins:

"It was a beauty, the Bluebird, a pleasure to drive…" [44].

1951 LRTL Tour with No. 1 at Archway. (Tom March/On Line Archive)

1951 LRTL Tour with No. 1 at Aldwych Station going south in the Kingsway Subway. (TLRS Collection)

1951 LRTL Tour at Streatham. (J H Meredith/TLRS Collection)

1951 LRTL Tour at Purley. (J H Meredith/TLRS Collection)

No. 1 at Penhall Road sidings prior to removal to Leeds. (G E Baddeley/TLRS Collection)

Chapter 7

LCC No. 1 becomes Leeds City Transport 301

Robert Morris

The story of the journey LCC No. 1 made from London to Leeds is closely linked to Victor Matterface, the Tramway Rolling Stock Engineer for Leeds City Transport. Matterface had previously held the post of Senior Technical Assistant (Trams and Trolleybuses) with London Transport, and had moved to Leeds in March 1948. Leeds City Transport had been struggling to renew its rolling stock; many of its older tramcars needed to be replaced, but it was proving difficult to find manufacturers willing to supply new cars. According to former Leeds City Transport engineer Geoffrey Hilditch, Leeds could have purchased fifty trams from Charles Roberts of Horbury which had been built using Sheffield 501 as the pattern, but insisted on retaining the nearside staircase and had to look elsewhere [1]. The poor state of the Leeds fleet caused problems and a system of overhauling trams during the off-peak hours of the day was introduced, enabling peak time services to continue [2].

The lack of available new cars led Leeds City Transport to purchase 37 second-hand vehicles from Southampton; these were of poor quality and could not be considered a long-term solution. A further six cars were purchased from Manchester in 1948, but the Leeds fleet still required an injection of modern and reliable cars if was to prosper during the 1950s [3]. As a former London Transport engineer, Victor Matterface was familiar with the London Felthams, and one was loaned to Leeds for trials. Matterface travelled to Charlton in late September 1949 where he personally collected Feltham 2099; this tramcar was well received in Leeds and later entered service as 501. At the same time, London was facing a shortage of buses, so in October 1949 Leeds City Transport hired out to London Transport seventeen AEC Regent diesel buses which were surplus to requirements due to staff shortages [4].

On 1 December 1949, A B Findlay became the new manager of Leeds City Transport and promptly endorsed the proposal to purchase a fleet of modern Feltham cars. In early 1950 the Leeds Transport Committee [5] resolved to authorise the expenditure of £73,600 to purchase 92 Feltham tramcars from the London Transport Executive [6]. The first Felthams arrived in Leeds in August of the same year and were sent to the Kirkstall Road works for overhaul and repair.

Leeds 301 with the bottom of the side panels cut away to clear the truck sides on sharp curves. (M J O'Connor/National Tramway Museum)

Then, on 18 November, two of the Feltham cars, LTE 2144 and LTE 2162 – which had cost Leeds £500 as a separate item - were destroyed by fire in London's Brixton Hill tram depot before they could be transferred. As a result, their trucks were salvaged and sent to Leeds as spares, but the loss of the two cars left Leeds with fewer trams than anticipated.

According to Eric Smith, the son of Tom Smith who served on the Leeds Transport Committee, Victor Matterface personally negotiated the transfer of LCC No. 1 in lieu of the fire-damaged Felthams, and the city's elected representatives only found out after the event [7]. Had Matterface not intervened and arranged for LCC No. 1 to enjoy a new lease of life in Leeds it may not have survived and would have perhaps become a half-forgotten footnote in the history of British trams. Records from London Transport reveal that LCC No. 1 was placed along with 529 other redundant tramcars on a list sent to the scrap trade with a quoted scrap value of £61. Although this indicates that there was little sentimental attachment to No. 1 in London, the delivery schedule shows that it was not expected to be disposed of until July–September 1952, almost two years into the scrappage schedule and amongst the final vehicles listed [8]. Richard Elliott, a former employee in the LCC drawing office who knew Matterface well, believed that LCC No. 1 owed its survival to the Brixton Hill fire and the Leeds Rolling Stock Engineer's enthusiasm for the experimental car [9].

LCC No. 1 arrived in Leeds on 19 June 1951 at 10pm, and entered service as Leeds 301 on 1 December of that year, operating from Chapeltown Depot. It had cost Leeds £500 to buy, with a further £300 being spent on overhaul costs [10]. The following alterations were made to the car before it ran:

- Bottom of side panels cut away to clear truck sides on sharp curves in Kirkstall Road works;
- Front and side service number boxes blanked out;
- Side via point boxes reduced in size to accommodate standard Leeds destination blinds;
- Bow collector fitted to roof and bow rope pulley wheels fitted on the roof ends;
- Blanking off panel fitted between trucks to fill gap where plough gear had been removed;
- Stair head drop windows replaced by sideways sliding type;
- Access handrail fitted to roof above stair head window;
- Lower deck electric bell pushes removed and replaced by rope operated electric buzzers;
- Curved glass window in front of platform doors replaced with steel panels;
- Air operated folding platform steps replaced with fixed steps;
- Metal-framed drop-down windscreens replaced with outward-opening wooden-framed windscreens with thicker frames.
- Remains of horizontal air operated windscreen wipers removed from top of windscreen frames;
- Conventional windscreen wipers fitted [11].

The cutaway side panels were meant to give extra clearance for the bogies on sharp curves, but the adaptation caused problems when the car entered service. According to Tom Steward, a Leeds car fitter at the time, this led to the fracturing of an air-brake pipe when the car negotiated a sharp curve [12]. An extra kink was fitted to the replacement pipe to allow the bogie to clear it. A recent discovery during the National Tramway Museum's restoration shows that the Leeds workshop changed the build-up of the bogie centre assembly, possibly in an attempt to get more clearance above the wheels on sharp curves [13].

At some point during LCC No. 1's time at Leeds, the wheels were re-tyred. This was revealed recently, when work on the trucks showed that the tyres had been manufactured by Firth Brown of Sheffield in 1952 [14]. The car received little cosmetic attention and was only repainted once, in February 1954, this time without lining and with larger dark blue fleet numbers. The trucks and lifeguards were repainted with red-brown oxide [15].

Tom Steward was the first Chapeltown employee to drive Leeds 301. He described it as a particular favourite of Oliver Anderson, the depot foreman. Anderson was keen to ensure that 301 worked regularly, despite the fact that some of the Leeds drivers were afraid of it "and would ask for a change-over on the slightest pretext. When this occurred, Oliver would sail out of the depot puffing fiercely on his pipe, emitting clouds of smoke like a tugboat under full steam and, telling the driver exactly what he thought of him, would treat that unhappy individual to a demonstration of driving guaranteed to turn him green with envy" [16].

A second bout of teething troubles occurred shortly afterwards when 301 began its first long duty journey and an electrical cable caught fire under the platform at the resistance end [17]. Unfortunately, the only wiring diagram was fixed on the cab door which was severely damaged by the fire [18].

A rare colour photograph of Leeds 301 in service, en route to Briggate. (M J O'Connor/National Tramway Museum)

After its initial repairs 301 was used for short journeys, but the electrical problems persisted with the circuit breakers blowing several times. Steward and his team tested the controller circuits but could find no faults. The problem was eventually found to have been caused by the handbrake chain swinging and catching the grids of the resistances located under one of the platforms, creating a short-circuit. According to Steward, this "could be blamed on poor track maintenance, because 301, being lightly sprung, would roll like anything over some of the dropped joints, especially near Reginald Terrace, thus causing the handbrake chain to flap about". The solution was to place some brackets underneath the platform and attach baffles made from insulating material between the handbrake chain and the resistances to prevent the short-circuit.

Re-entering service, the car derailed on Stainbeck Lane spur in Chapeltown on its first day; but the electrical and mechanical problems were eventually resolved and, as Steward noted, "it was a great success and we had no further trouble with 301. In fact, some of our drivers even grew to like it" [19]. After the initial setbacks had been dealt with by the workshop, 301 settled down to operating from Chapeltown Depot on services 1, 2 and 3 from Lawnswood to Roundhay via Harehills, the only time it was used regularly on all-day service [20]. Apart from routine maintenance, the other major alterations to the car occurred in March 1953 when bow rope tubes were fitted as replacements for the original pulley wheels [21], and in May 1956 when Peters type air wheel brake operating sets from a withdrawn Horsfield car were fitted [22]. The former air wheel brake operating equipment on top of the controllers was retained but not used, and the EMB interlock was kept to operate the air/magnetic track brake.

An uncorroborated story mentions that 301 ran away whilst being moved in the depot yard because the driver's air brake valve worked the opposite way round to all the other air braked cars in the Leeds fleet. The valve handle in 301 had to be pulled towards the driver to apply the brakes, unlike the other cars where the handle was pushed away from the driver [23].

According to the May 1956 issue of Leeds and District Transport News, the air wheel brake equipment "was operated in the opposite way to that on all other Leeds trams in recent years (i.e. 'on' towards the driver and 'off' away from him on No. 301), and this has been one of the reasons for the car's restricted use since it came to Leeds. Since being re-equipped, the car has become one of the most popular cars with the drivers, and for the first time since the closure of the Chapeltown Depot to service cars has been used regularly for all-day duties. However, after about a week, No. 301 again fell into disuse again (sic), and by the end of May was only going into service for peak hour duties, mainly on the Hunslet

Leeds 301 in the city centre on the Lawnswood 1 route, 7 April 1954.
(R B Parr/National Tramway Museum)

The fire brigade attending a fire on Leeds 301 associated with the electrical wiring. Date unknown, but after the addition of a rope tube to replace the original pulley wheel. (A D Packer)

Leeds 301 at Parkside level crossing during the Light Railway Transport League tour of the Leeds system, 30 March 1952. (R B Parr/National Tramway Museum)

*Leeds 301 and 278 on the Moortown Circular route, 10 May 1956.
(A K Terry/National Tramway Museum)*

The lower deck of Leeds 301 on Dewsbury Road, 28 May 1957. The cramped standing area for the conductor is clearly visible. (M J O'Connor/National Tramway Museum)

route" [24]. During this period 301 suffered at least one other fire associated with the resistances or wiring, with the fire brigade having to attend.

The fortunes of 301 were intertwined with those of the Leeds tramway system as a whole, and its failure to carry out its duties to its full potential was partly due to the declining tram service as the 1950s progressed. As a unique, experimental vehicle, the car was of great interest to enthusiasts and on 30 March 1952 it was used for a Light Railway Transport League tour of the Leeds system, covering the routes to Gipton, Middleton and Belle Isle [25]. On 7 January 1955, 301 was used for a journey along Dewsbury Road, the first time that it had been spotted in service on routes other than Lawnswood, Moortown and Roundhay [26]. When the Chapeltown Depot was closed in April 1955, the car was transferred to Swinegate where it was mainly employed for peak hour services [27].

In November 1955, 301 was noted in service on the 25 Hunslet route [28], and in 1956 the car was seen regularly working the Moortown, Roundhay and Dewsbury Road services [29]. Although a popular car with enthusiasts, its reputation amongst the Leeds City Transport staff was more mixed. As noted earlier, some drivers were afraid of it and had to be coaxed into taking it on the road [30]. Chris Thornburn, a former Leeds conductor, recalls that after the May 1956 air brake control replacement, 301 should have been more popular with the drivers "and indeed, for a week or so, it was well-used, but then reverted to the former peak hour only appearances. Maybe continuing driver prejudice, the shed staff realizing there was no point in rostering it only for drivers to refuse it – at peak times they could say 'You'll have to tek it, there's nowt else!'. Or perhaps a general antipathy, or even policy against anything different or non-standard" [31]. Others such as Victor Matterface and Oliver Anderson could appreciate 301, and former Leeds engineer Geoffrey Hilditch later described it as "without a doubt the best tram we bought... a vehicle that could show any Feltham a clean pair of wheels. What a pity it was that that design was never multiplied" [32].

During the summer holidays Leeds City Transport frequently suffered from staff shortages, so the Transport Committee decided to enrol local students on vacation to work as conductors on the trams and buses. After a brief week-long training period a new set of fifteen student recruits started each week, amongst them Leeds Transport Committee Deputy Chairman Tom Smith's son, Eric [33]. Eric Smith was initially trained as a bus conductor, but after his father used his influence, he was told to report to Swinegate Depot for special instruction in Tramcar Braking. On 24 July 1956 he was sent to work on 301 and described his working day:

"What a surprise! We found we'd been given car 301. The 'board' indicated no journeys to Dewsbury Road itself but a return trip from Briggate to Harehills, followed by two return trips – Briggate to Chapeltown – and ending with Briggate to Moortown 2 circular. I was looking forward to that – a ride along the express (reserved) tracks in Princes Avenue and Roundhay Road. My driver didn't seem too happy. Traffic might have been bad. Or he might have been dawdling. Anyway, when it came to the final trip of the 2 circular, the inspector regulating departures from the famous Briggate barriers decided that we were so late that the last trip would be, yet again, to Chapeltown only. I found the car a little difficult from the conductor's point of view. There wasn't a suitable space to stand, other than to disappear into the cab, which circumstances of people boarding forced you to do. You couldn't flatten yourself against the cab door... it was on some sort of ball-bearing catch and swung both ways" [34].

Leeds 301 at the rear of Swinegate Depot on 2 July 1957, two months before it was withdrawn from service in the city. (M J O'Connor/National Tramway Museum)

Another student recruit was Chris Thornburn, and in July 1955 he was offered overtime working as a conductor on 'Children's Day', a popular event in Roundhay Park. This was a busy day for the Transport Department "when even the grumpiest old driver who normally wouldn't touch overtime would condescend to do a bit 'for t' sake o' t'kids' ".

Mr Thornburn continued: "At 5pm I reported back to Swinegate, and found myself allocated to, of all things, '301 car track 2'! With a shedman driving me – who, unlike many of the regular tram crew, knew how to handle 301 and wasn't afraid of it. And we spent a pleasant evening doing half-a-dozen round trips taking people home from the park. One thing struck me straightaway was that the platform layout is somewhat cramped, and the conductor tends to be in the way of passengers boarding and alighting; the solution was to stand halfway into the back cab, from where you could still see what was happening on the platform step. On one trip, going up Princes Avenue reservation with nobody on board, Bill beckoned me forward – 'D'you want to have a go?'. Naturally, I didn't have to be asked twice! So 301 became the first tram I conducted on my own, and the first I ever drove. It might even have run to Middleton had it not been for a timid driver. Some time after my Children's Day experience, I was allocated a Middleton duty; looked it up on the going-out board, to my surprise saw 301 as the rostered car! Wow, this is an 'Unusual Working' for the record! Set up my destinations, and waited for the driver to appear – whose first words were 'Ah'm not tekkin' that b****y thing!' and carted me off to Showboat 164, instantly dashing my initial euphoria..." [35].

During late 1956 and 1957, 301 worked occasional duties to Balm Road during evening peak hours [36] and on 2 May 1957 it became the last tramcar to use Vicar Lane in a southerly direction [37].

In September 1957 the Moortown and Dewsbury Road services were closed and 301 was withdrawn from service in Leeds [38]. The gradual running down of the Leeds tram system had begun in 1951 when a Transport Department deficit of £139,393 was recorded and the incoming Conservative group on the City Council set up an all-party Special Investigations Committee to examine options for reorganisation and financial economies [39]. It was discovered that the trams had for many years been subsidising other Corporation activities due to incorrect cost allocations, and in January 1953 a short section of track at Stanningley became the first part of the system to be abandoned. At the same time, an increase in the cost of traction current of £35,000 per year added to the financial pressures. The deficit had risen to £318,430 by March 1952 and more heavy losses were expected in 1953, causing the opposition Labour group to start a campaign against the trams [40].

Although the Labour group accelerated the closure of the tram system when it took power in May 1953, the decision to phase out trams was not taken on party political lines. Alderman Mather, a former

Leeds 301 on the Moortown 2 route, 9 October 1954. (M J O'Connor/National Tramway Museum)

The upper deck of Leeds 301 on 28 July 1957. (M J O'Connor/National Tramway Museum)

Conservative member of the Transport Committee who was in favour of retaining and expanding the tramway, pointed out that this "was not really a Party matter as there were many Socialists who agreed with me, and no small number of Tories who agreed with Alderman Rafferty" (that trams should be replaced with buses) [41]. *The Yorkshire Evening News* ran a prominent newspaper campaign against the trams [42] and the bus lobby in the end proved victorious – the last Leeds tram ran in November 1959.

Leeds 301 was spared the fate of most of the City Transport fleet and survived to become a much admired museum exhibit. It was announced on 18 March 1957 that the car was to be presented to the British Transport Commission, although it is not known when the decision was originally made or if Victor Matterface had any part in its preservation. *Leeds Transport News* recorded the departure of 301 to London in its December 1957 issue:

"The Pickford's lorry ULY 923 left Swinegate Depot with 301's body on a trailer at 4.15pm on 26th November, and was parked in South Accommodation Road Goods Yard forecourt that night, leaving Leeds via York Road and Cross Gates on the following morning. During the 27th November, the load travelled via the Great

Leeds 301 and 600 in service on 14 June 1957; these experimental trams would later be reunited at the National Tramway Museum. (A K Terry/National Tramway Museum)

North Road as far as Doncaster, and was stored that night at the British Road Services Depot there. On the following day, Thursday, No. 301 was taken (again via the Great North Road) as far as Kate's Cabin, near Stamford, and the journey was completed on the Friday afternoon when the tram was delivered to Charlton. The body was unloaded from the trailer on Monday, 2nd December, and it now stands on its bogies again in the wired compound inside the Works" [43].

The tramcar was now back in its home city of London. In 1959 it was moved from Charlton Works to Clapham; on May 29 1963 the large exhibits section of the Museum of British Transport at Clapham opened to the public, and the next chapter of the tramcar's story began.

Leeds 301 on the Leeds ring road at Halton, at the start of its journey to London on the morning of 27 November 1957. (A K Terry/National Tramway Museum)

Chapter 8
Preservation
Hannah Bale

The British Transport Commission and its Consultative Panel

Before 1948, the principal organisation preserving a national collection of historic vehicles was the Railway Museum at York, which naturally focused solely on railways. With the 1947 proposal to nationalise the railways, however, came the realisation that more needed to be done to preserve a wider selection of historic forms of transport, under one organisation.

The British Transport Commission (BTC) was formed in that year, and in a letter to *The Times* of 31 December 1947, Sir Cyril Hurcomb (later Lord Hurcomb), President of the Stephenson Locomotive Society (SLS) and the Commission's Chairman, stated that it was concerned with all forms of transport. He also suggested that transport artefacts could be displayed in a central location for the enjoyment of the nation.

On 19 March 1948, Sir Cyril met with Mr R A Riddles, the Commission's Engineering Manager, and Mr W O Skeat, who had already taken an active role in locomotive preservation and who advocated the foundation of a national transport museum, and the Consultative Panel for the Preservation of British Transport Relics, which was to include representatives from a large number of prominent societies, was formed [1] [2].

Interest in the Panel was immediate and, in response to Sir Cyril's suggestion, work began to compile a list of all locomotives and rolling stock deemed worthy of preservation. Although the primary focus was on the railways, other items, including tramcars, were included. Other like-minded societies followed suit, producing their own recommendations, and in 1951 the completed lists were sent to the British Transport Commission. Mr John Scholes, newly-appointed Curator of Historic Relics, examined them, and in the same year published *The Preservation of Relics and Records*, in which he outlined what the Commission would preserve and what classification those items would fall into. 'Relics', which concern us here, were items, including rolling stock, deemed to have antique value or interest [3]. Today we prefer to use other terms to describe them, such as 'historic objects' or 'artefacts'.

The Museum of British Transport, Clapham

In the Commission's early days it was agreed that the preservation of these relics could involve their display in a museum setting for the general public to visit [4]. The London Transport Executive intended to keep a small number of tramcars as historic specimens, but did not have anywhere suitable to house them [5]. These, together with the Transport Executive's collection of omnibuses and the collections of London-based railway and canal companies, formed a large portion of the vehicles to be displayed in the proposed museum [6]. In 1955, the Light Railway Transport League (LRTL) produced their own list of tramway items to be preserved – the only list to survive – and in the following year, Mr Scholes had the first vehicles, including tramcars, secured for preservation [7].

Due to issues with finance, it was decided to undertake the opening of the museum in stages [8], placing many of the large artefacts in storage until an appropriate, permanent site had been found for their display [9].

Disaster was about to strike, but as a result the Consultative Panel was to become an even stronger force. In 1957, without the permission of Mr Scholes, some of the items earmarked for preservation were scrapped. Interested societies complained about the scrapping in the national press, and representatives were appointed to meet with Sir Brian Robertson, the new Chairman of the British Transport Commission, to resolve the issue. Sir Brian recommended that an advisory body of representatives be formed, covering all areas of transport and presenting to Mr Scholes those artefacts and vehicles which should be preserved [10].

The Panel, which now included the Tramway Museum Society (TMS) among its component societies, gathered for the first of what would become quarterly meetings in July 1958, with Mr Scholes in attendance. The individual societies formed new lists of the items they believed suitable for preservation [11], having split for this purpose into sub-committees, each concentrating on a particular mode of transport. One of these was the Tramway Sub-Committee. The completed lists were presented to the British Transport Commission, through Mr Scholes, for consideration; the Commission also added their own suggestions [12].

Mr Scholes eventually secured the Museum of British Transport site at Clapham. The site had been home to a tram depot and, more recently, to a bus garage; it had become available after having been closed by London Transport, and would be suitable as a permanent base for the collection. It would house those collections relating directly to the capital, plus disparate collections at that time held elsewhere in England [13], while also acting as the central point for two other museums: the Railway Museum at York, and a museum in Edinburgh for Scottish collections.

On 29 March 1961 the Exhibition of Small Objects was opened to the public, and on 28 March 1963 the Exhibition of Large Relics, containing the locomotives, buses and tramcars, followed [14]. Tracks were re-installed to carry the locomotives and tramcars, the tram tracks being arranged into a 'Tramway Avenue' on the north side of the building. Overhead wires were also installed for the tramcars and trolleybuses to give a more authentic appearance [15].

LCC No. 1 at Clapham

Mr Scholes had established contact with the Museum Committee of the LRTL in 1955. Later that year the Museum Committee became the Tramway Museum Society. At that time the Museum of British Transport had only three tramcars, all previously owned by London Transport, in its collection. By working with the LRTL, Mr Scholes hoped to secure additional tramcars to better illustrate the history of trams in Britain [16]. In the following year, he also contacted the Tramway and Light Railway Society, with the aim of introducing models to the exhibition. It was with the aid of these two societies that Mr Scholes was able to secure the preservation of LCC No. 1, then Leeds 301, when it was offered to the Commission by Leeds City Transport [17]. No. 1 was considered an important vehicle to preserve due to its importance in modern tramcar development, particularly regarding improved riding qualities [18].

In the last week of November 1959, LCC No. 1 arrived, along with Sheffield 342, recently withdrawn after some fifty years of service in Sheffield [19], at the Museum of British Transport from the storage depot in Charlton [20].

Two years later, Mr J H Price, of the TMS, and the Tramway Sub-Committee discussed and wrote a paper on the history of British trams from 1914 to the 1960s, reinforcing the argument for preserving certain

Lineup of trams at the Museum of Transport, Clapham, with LCC No. 1, still in its Leeds livery, in the centre. (© Ray Reed/RCTS Photographic Archive)

tramcars, including LCC No. 1. Due to the lack of tramcar development in the 1920s and subsequent design work undertaken by the LCC and English Electric in the 1930s to bring tramcars into the modern era (as seen previously), No. 1, along with the modern composite-bodied Blackpool cars, was considered a defining influence in the history of the British tram. Tramcars built after No. 1 and the Blackpool cars took on characteristics of these experimental cars, and all the features used on modern tramcars were to be found on No. 1, such as the all-metal bodywork, improved lighting and the white-painted ceiling, reflecting more light [21].

Following further tramcar acquisitions between 1961 and 1963, the Tramway Sub-Committee's next task was to compile a report on the restoration of each vehicle. Some of them had been extensively rebuilt and it was considered impossible to return them to their original state, so it was decided that scale models would be created to represent the tramcars as they had been when new.

By this time, however, changes at Clapham had cast doubt over the future of the collection.

In 1963, following the Transport Act of 1962, the British Transport Commission was dissolved and the Museum of British Transport and its collection was placed under the control of the new British Railways Board. Unlike the Commission, which was dedicated to the preservation of transport history, the Board wanted to be forward-looking and present a modern image. A transport museum did not fit these criteria. A concerned Consultative Panel decided after considering the options that the best way to ensure the museum's continuance and protect the collections would be to approach the Government, through the Department of Education and Science, and ask it to take responsibility for all three transport museums [22].

The museum at Clapham cost the British Railways Board around £90,000 a year, predominantly in staff wages and the restoration programme. The entrance fee was not enough to cover this amount, despite the museum's popularity. Although such a financial deficit was expected in any staff-run museum, Dr Beeching's move to improve British Railways' finances brought the drain the museums represented on the Board's resources to his attention. It was Dr Beeching who put forward the suggestion that the Government should care for the museums [23], but on 9 March 1964, the national press revealed that the Government had rejected the proposal.

In the first days after the decision was made, it was hoped that public opinion would help sway the Government. A General Election would soon take place, and the political decision would be one of the first the new Government would pass. The campaign to save the museum was taken up individually by every society which made up the Consultative Panel, and J H Price, in an article on the situation published in *Modern Tramway*, encouraged readers to write to their MP. Letters to local papers and the transport press were also encouraged, to bring the situation to wider public attention and hopefully bring in returning and new visitors [24].

By 1965, the tramcars in the collection, with models and photographs, effectively represented the history of the tram in Britain. The Museum Board hoped to develop the collections further [25], but without financial assistance they were forced to reduce outlay in any way they could. Many exhibits were returned to storage, restoration projects, particularly large ones, were put on hold, and in addition it was realised that some of the collection would have to be disposed of to reduce the burden of care. The items to be retained would be those judged to be the greatest draw for visitors, irrespective of condition or historical value [26]. Only exemplary items would be kept in the museum's collection. The Consultative Panels had been re-examining all the exhibits to determine each item's importance and exhibition appeal [27].

Finances were not the only factor to affect the museum and its collection. The original purpose of the Museum of British Transport was to show a full history of all public transport vehicles, but the transfer to British Railways, with assistance from London Transport, changed this outlook. British Railways saw locomotives, particularly the famous engines, as the priority for preservation and immediate display. Some exhibits held at York, such as their City & South London Railway coach, were also thought to be more suited to display in London. With these aims in mind, the first 'relics' to be considered for disposal were the tramcars [28].

The cars were divided into three groups – historical significance to London, visitor interest and antiquity - and each had provision made for its future. London Transport, who provided financial support to the museum, felt that tramcars which had particular significance in the history of London's tramways should continue to be displayed. They selected three trams: ex-West Ham 290, Class E/1 car 1025 and Feltham car MET 355. The British Railways Board were also willing to keep those tramcars most popular with visitors. Mr Scholes believed these to be the horse trams and early electric cars, which were the oldest and which had particular antique appeal; another five tramcars were chosen to fill this role. Douglas Head Marine Drive No. 1 would remain permanently and Llandudno No. 6 indefinitely; Blackpool No. 1 would remain for the time agreed by the TMS, under whose ownership it fell, but would later return to their museum at Crich; and Douglas 14 and Chesterfield 8 would also remain, but would be offered to other museums if space was needed for other exhibits [29].

Mr Scholes, after consultation with the Sub-Committee, chose a number of trams to be considered for disposal to another museum. Grimsby and Immingham 14 was offered to the TMS. Snowbroom 022, previously LCC 106, which had been converted and renumbered in 1925, and LCC No. 1, both in the ownership of London Transport, were also marked for disposal [30], along with Glasgow 1392 which was to replace Newcastle 102 at Beaulieu, the latter going to the museum at Crich, and Sheffield 342, which went to Beamish Museum [31]. Grimsby and Immingham 26, at that time held in storage, was to remain until it was needed for a proposed northern folk museum [32].

On 31 October 1966, during a meeting held at the Museum of British Transport, J H Price confirmed that LCC No. 1 was one of the tramcars in the collection marked for disposal. It had been offered to the TMS on permanent loan, to be housed, restored and exhibited by them until they no longer had need of it, when it was to be given back to the British Railways Board; but the TMS had stated that they would only take the tramcar if it was offered outright [33]. They were keen to take No. 1, both to better represent London trams at Crich and to provide another tramcar for TMS members and visitors to enjoy [34]. The TMS, however, faced issues with logistics and housing, as there wasn't room to store the tramcar at that time unless something else was moved out of the Depots. It therefore asked for six months to prepare for the move, provide housing and run an appeal for funding to pay for transport, accommodation and eventual restoration [35].

One possible solution under discussion was for Acton Works to take the bogies for overhaul work and apprentice training [36]. It also seems that at some point, an enthusiast had offered to pay for the tram's restoration, provided it was returned to its blue livery. However, it was feared that if this offer was accepted, less money would be forthcoming for transport and housing. Donors, it was argued, would also want to have a say in the livery colours, and whether the tramcar should be returned to the LCC, London Transport or Leeds part of its history [37].

An appeal was launched by the TMS in August 1967, to raise funds for the removal of LCC No. 1 and Snowbroom 022. Developments at the Clapham Museum then gave them an apparent respite of a few months to raise more money and better prepare for the removals [38]. These developments so changed the circumstances, however, that a few months turned into over five years.

There were two reasons for the delay. Firstly, British Railways changed its mind over bringing the locomotive 'Evening Star' out of storage for display at Clapham. In the short term this made things easier for the TMS, providing them with more time to raise funds, organise haulage and insurance, and overcome the issue that LCC No. 1 was blocked in by Glasgow 1392. It also gave them time to concentrate on the removal of Snowbroom 022 [39].

Secondly, The Transport Act of 1962, which appeared to supersede the previous arrangement made between the Museum of British Transport and the TMS, could have prevented LCC No. 1 coming to Crich at all. The Act stated that responsibility for the preservation of the 'relics' rested with the Railways Board [40]. It also gave permission for the Board to make arrangements for the collection as it saw fit, by loan, sale or gift, for temporary or permanent custody to any other appropriate collection. The Board could also dispose of items no longer considered necessary to the museum's collection, "by gift or sale or otherwise", although nothing could be disposed of without the matter first being put before the appropriate Sub-Committee with one months' notice. The first offer would be to the Railways Board and transfer of ownership was free [41]. Newspaper articles at the time announced that all exhibits held at Clapham would be placed in storage to await further developments [42].

The Transport Act also provided for the transfer of the railway exhibits held at Clapham to the museum at York, with Royal assent. Opposition to the move came from 17 London Borough Councils, who urged the Greater London Council (GLC) to open a new London transport museum for the artefacts relating to the capital, including tramcars, which would otherwise be placed in store [43]. A report in *Norwood News* stated that the GLC was considering the proposal, and the new museum was to be built in the grounds of the Crystal Palace. It would house all the London Transport exhibits at that time held by the Clapham museum [44].

The proposal of a London transport museum was also backed by the Transport Trust, who believed a new museum was entirely feasible, at no greater cost than the one already open in York. By 1970, the campaign for the new museum was nearing success. The removal of exhibits to York was suspended while the scheme for the London museum was fully worked out and submitted by the Transport Trust. They proposed to use the east wing of Crystal Palace Station, which over two floors would house railway exhibits and road vehicles. The buses would be able to run on the GLC-owned Crystal Palace racing circuit [45].

The scheme was scrapped, however, due to the sum the British Railways Board asked for the land, preventing its implementation for five years. In December 1971, the Government announced that the railway exhibits in the Clapham museum would be transferred to the ownership of the Department of Education and Science and moved to York. The Department had first option on all these railway exhibits, with other nationalised boards having first option on all other items associated with them. Under this stipulation, London Transport could claim not only the tramcars, buses and trolleybuses it had contributed to the museum, but also Feltham MET 355, the Tilling-Stevens bus chassis and LCC No. 1, even though they had been donated by other bodies. This would place No. 1 in the ownership of London Transport.

By this time, London Transport was also considering opening its own museum for all transport relating to London, in a similar style to the Museum of British Transport. In the meantime, the Museum of British Transport would be slowly run down, with reduced staff and opening hours until it finally closed [46].

Although official legislation enabled London Transport to take LCC No. 1, as well as other items for their own future museum, they chose to stand by their previous offer to donate the tramcar to Crich [47]. In August 1972, London Transport clarified their situation to the TMS, stating that they would only take No. 1 if the TMS could not, or no longer wished to, take it. The TMS Board immediately confirmed their interest, and new arrangements were put in place to finance and move the tramcar as soon as the exit was clear. This was finally made possible when Glasgow 1392 was removed by Glasgow Museum of Transport on 4 November 1972 [48].

LCC No. 1 comes to Crich

In December 1972, LCC No. 1 finally made its way to Crich from Clapham. The move was arranged by Mr David Packer. On 30 November, the underslung gear, including the lifeguards and pilot beams, were removed and the motor cables disconnected and labelled ready for transport. Under the eye of Mr T Courtney, of London Transport, No. 1 was towed out of the Clapham museum building. On 1 December, Elliott (Hauliers), of Rufford, Yorkshire, arrived at Clapham to prepare for the tramcar's removal. No. 1 was lifted on its rails with three dollies – beams with three small wheels – clamped under the rails. The following day it was loaded onto Elliott's lorry [49].

On its way out of London, No. 1 passed its old home at Telford Avenue Depot, a happy coincidence as the route was only planned in that direction to avoid a low bridge at Balham Station [50]. Disaster almost struck when, at Ealing Common, the lorry trailer came close to breaking under the tramcar's 21 ton weight, but the journey thereafter was fortunately uneventful [51]. Following an overnight stop in a lay-by on the A6, eight miles south of Derby, LCC No. 1 arrived at Crich on Monday 4 December, and was unloaded the next day [52].

Stabilisation and conservation

At Crich, No. 1 was taken to the Workshop to have its lifeguards, and other items which had been removed, refitted. It also received a deep clean [53]. Two trolley booms and two trolley bases, which along with other spare parts had accompanied it on its transfer from Leeds to London, did not travel with it to Crich; a bow collector which had been attached while in service in Leeds did arrive, although the tram's bow collector fittings had been removed [54].

Examinations of LCC No. 1 revealed the extent of the work carried out during its service in Leeds, and showed that no major work had been done since then.

At this point, it was not known which livery No. 1 would run in. Indeed, long-term plans for its future had not been decided, partly because the final move had been so sudden [55]. Initially, a long-term plan

No. 1 being loaded onto Elliott (Hauliers) Ltd's truck outside the Museum of Transport, Clapham, before departure for Crich, 2 December 1972. (National Tramway Museum)

On Poynders Road, Clapham, 2 December 1972. (National Tramway Museum)

Examining the trailer on breaking down at Ealing Common, 3 December 1972. (National Tramway Museum)

On the M1 Motorway heading for the Midlands, 3 December 1972. (National Tramway Museum)

was put forward for the tramcar to be given enough attention to allow it to run as Leeds 301. Due to other projects already under way or about to be undertaken, the only short-term work proposed for the tramcar was to carry out necessary mechanical and electrical work before a later repaint [56].

It appears that little funding was left over for care of the tramcar, and even the minor repair works which were proposed when it arrived were not carried out [57].

By April 1975, No. 1 had taken up residence in Depot 3. No further work had been carried out and the tramcar remained as it had been on leaving the Clapham museum. It was soon realised that it would need a complete overhaul to place it within the running fleet [58]. However, there was no move to make it serviceable again; it was not included in the budget for tramcar works in 1976-1977, as it was considered not to be needed for use in service [59], presumably because it had entered static display.

In 1976, a Job Creation Programme, funded by a grant application, was in place, allowing a small workforce to restore up to five tramcars [60]. Under this scheme, in March 1978, LCC No. 1 was repainted in London Transport livery before being placed on display in the Exhibition Hall [61]. While the repaint was carried out, other works were also undertaken; various fittings were removed or altered to return the tramcar to a closer representation of its condition while in service in London. The front service number blind boxes were recreated, but in the form of simple, plain wooden frames, rather than to the original 1930s design. Non-operational side destination displays were fitted at the same time [62].

As no work had been carried out on the equipment and traction wiring since No. 1 left Leeds, these showed a mixture of original London material and Leeds modifications. The electrical and air equipment had been modified or replaced while in Leeds and was typical of that city's trams [63]. Sometime before coming to Crich, all the bow collector equipment including the body fittings, with the exception of the plank - or trolley bar – had been removed [64] [65]. Consideration was given to the replacement of the trolley bases and equipment in 1976. It was suggested that new trolley bases could be built, either to manufacturer Brecknell Willis' original drawings if they could be found, or by converting a "standard BW base by dividing one compression spring to make two short ones, and purchasing the tension springs" [66]. This work was not ultimately carried out and new trolley bases have finally been created during the present restoration.

The 1976 examination revealed evidence of the removal of several items connected with the controllers and operating of the tramcar, carried out in Leeds. An engineering drawing – *OK 37B Controller schematic: 4 motors (LCC 1)* – showed that while in London the tramcar contained a plough-trolley changeover switch or switches. This was not present, nor was there any clear location for its position, but there was definite evidence that Leeds had removed the negative end switchgear and fuses; the interlock circuit breaker may have been removed at the same time [67].

In early 1991, enquiry was made into the replacement and updating of the upholstery and moquette. That on the lower deck would be easy to replace, but that on the upper deck is of a different pattern and would have had to be specially made. It was agreed the work would be included in the summer budget; the expense of the upper deck material would be justified by enough of it being ordered to fit out LCC No. 1 and at least two other vehicles [68].

Ultimately, the work was not carried out and the moquette and upholstery was left in its original state.

It was to be another 12 years before LCC No. 1 would be again considered for conservation work. During the London County Council Tramways Trust (LCCTT) Annual General Meeting of 2003, supporters of No. 1 put forward a motion to create a fund for the restoration of the tramcar to operational condition [69]. The proposal was passed and communicated to the LCCTT membership in their newsletter, *The Belvedere Echo*, calling for support for the project, with the fund opening in November 2003. At this point there was no time scale for the restoration's commencement or completion, but the LCCTT saw that funds from a wider pool would need to be drawn upon. A detailed discussion would also have to be undertaken between the LCCTT and the Tramway Museum Society, to decide what work would be needed to bring No. 1 back to "original LCC condition as a working exhibit in the National Collection." [70] The work was to be carried out on site by the Tramway Museum Society's Restoration Workshop.

Progress was slow, however. Although donations had been received for the project since the fund was created over two years previously and some £6,500 had been raised, with donations continuing to come in, by early 2006 there was still no formal fundraising for the restoration.

At this point, the LCCTT made a formal offer to the TMS Board of Management "to raise funds to support

The upper deck seating in 2012, showing the deterioration of the moquette; in some areas there was a complete loss of fabric, revealing the seat padding below. (D J Heeley)

the eventual restoration of LCC 1." [71] It was also declared that some monies would be granted from the Trust's general funds. The size of this monetary support, however, would depend on the expenses incurred by the London United Tramways 159 restoration which was then being carried out. As previously stated by the LCCTT, there was no specification of a timescale for the work to begin. However, the requirements of the final restoration had become more precise, with the desire the tramcar be "returned to a condition in which it existed between construction in 1932 and the formation of the LPTB in July 1933." [72] The LCCTT's offer of funds for the restoration of the tramcar was accepted by the Board, but LCC No. 1 would still have to wait until there was time and physical space in the Workshop for its restoration to begin [73].

A year later, by April 2007, the LCCTT's fundraising efforts had secured £7,500, with more donations being given, and £40,000 had been designated for the restoration from the Trust's General Funds, an amount which could be adjusted at any time by the LCCTT depending on any circumstances that might change or arise. How much the restoration would finally cost, and when it would begin, was still uncertain, the LUT No. 159 restoration project being still ongoing [74].

Meanwhile a plan had been developed to preserve the original interior furnishings rather than replacing them with new. In May 2008 a thorough examination of the moquette and upholstery was carried out and detailed recommendations were made, but again the seating was left in its deteriorated condition until the present restoration [75].

Preparations for restoration: the 2012 Condition Assessment

Over the next four years, little or no conservation work was carried out. During this period, the Museum's Curator also raised queries as to whether restoration was the right course of action for the tramcar, and discussions on the matter continued. Then, on 10 November 2012, during the 58th meeting of the National Tramway Museum Tramcar Conservation Committee, a non-intrusive inspection was requested, to assess No. 1's present condition and to help determine whether it should be conserved or restored. By this time, due to the limited work carried out previously, which had been largely confined to removing Leeds modifications and replacing London features, the tramcar was neither in a Leeds nor a London Passenger Transport Board condition [76].

LCC No. 1 was suffering in other ways as well. The design of the body, coupled with natural deterioration, was causing damage. The sealants and material coatings had deteriorated, allowing water to penetrate the body structure. There was also a significant amount of corrosion; the aluminium and steel components were reacting against each other, a process known as electrolytic corrosion, which was occurring due to high moisture levels. This was causing the tramcar body to distort, forcing the window pillars outwards; a development of particular concern which needed monitoring, as further distortion could cause sharp edges to protrude, posing a potential risk to passing visitors. Holding the tramcar in dry storage conditions would delay the development of further deterioration and distortion [77].

The 2012 examination revealed that the repaint into London Transport livery had been carried out with the minimum of preparation, giving a poor finish and leaving the paint in a poor condition [78]. The roof appeared to be relatively sound with little or no visible distortion to the plywood construction, although the paintwork and possibly the main canvas was crazed. The distortion to the structure began in the frame sides, appearing in the finish above the upper deck windows. Here the paintwork was lifting from the frame, cracking horizontally in small, short fractures with longer, deeper fractures elsewhere. Some small patches of paint had been lost; paint was also failing on the dashes, where the red London Transport livery colour had not adhered properly to that applied by Leeds and was rippling [79]. Corrosion was also occurring on the main pillars and the aluminium/steel join under the upper deck windows. The fixings for the metal strip covering the aluminium/steel interface, dating from the tramcar's time in London and Leeds, was failing, causing the strip to distort. In places there was also heavy corrosion behind this strip. Without removing the strip, it was difficult at that stage to determine the integrity of the structure [80].

Distortion was also apparent beneath the upper deck windows, where a strip of metal was lifting away from the body. On all the main pillars there was evidence that the main frame was delaminating, due to water ingress while No. 1 was in operation. This was forcing the main frame to distort, pushing the lower deck window pillar caps off, particularly above the north platform and the northwest side of the body. The upper deck advert strip was variable in size, suggesting it too had lifted and been filled in behind when the repaint had been carried out. Basic wooden cappings were found on the external frames of the lower saloon, and it was believed that they had been added, again at the time of the repaint, to disguise corrosion. The main underframe was covered in surface rust and general dust, dirt and other contaminants accumulated during the tram's service in London and Leeds [81].

It was believed that, with some investment, the condition of the interior could be stabilised, although many of the fixtures and fittings were damaged beyond recovery [82]. However, apart from the information gleaned during the 2012 examination, the condition of LCC No. 1's bodywork remained largely unknown.

The tramcar's truck and brake mechanisms were all complete, but past examinations had shown that the brake linkages were seizing and the brake system was at a point where it was considered no longer usable. Having worked with similar Electro-Mechanical Brake Co. Ltd (EMB) trucks in the past, the Workshop team had discovered that the main brake beams were prone to seizing on their pivot arms [83].

The roof in 2012, the coating of which had become badly cracked and crazed. (D J Heeley)

The distortion of the main frame can be clearly seen with the outer panel of the window frame buckling outwards. (D J Heeley)

The interior had received little or no attention since its time in Leeds. The interior panels were in poor condition, with heavy crazing and peeling on the woodwork. This deterioration of the paintwork was evident throughout, including on the ceiling and the wall panels above the windows. The panels below the windows were in a better condition, with less crazing, but still showed cracking and loss of paint in areas, along edges and panel seams. In the cab, the paintwork was also in various stages of deterioration. That around the windows was in a fair condition, but the ceiling was much worse, with evidence of mould damage [84]. Corrosion, in the form of white spotting or mottling on the surface of the metal, was also found on the interior chrome fittings [85]. While the tramcar was in Leeds, repairs to the floor covering had been carried out as the original, a blue material, had worn or become damaged. These repairs, in brown material, were failing, in places having shrunk along the joints; the material had also become brittle as it had dried out. In the cab and on part of the lower deck, the floor covering was also missing [86], revealing the wooden floor boards below [87].

Work had been carried out on the driving equipment in the cab while No. 1 was in service in Leeds. The plough gear had been removed along with the 'negative' side electrical equipment and fusing. This was evident in the fuse board under the north staircase. Following its removal from Leeds, it did not appear that any further work had been carried out and the control equipment seemed to be complete. However, the electrical equipment and wiring still in situ was deemed unusable.

The braking system was still in situ and complete, but was not the original LCC system. This had been modified in Leeds, with a standard Horsfield Maley & Taunton motorman's valve fitted alongside the EMB interlock on top of the controller.

Cosmetically, the controllers and equipment were in a poor condition, with peeling or missing paintwork. The equipment for the powered doors was present but the folding step had been removed in Leeds [88].

While in Leeds, a blanking panel had also been fitted over the opening beneath the tramcar where the conduit equipment had previously been. The panel had been removed, either after No. 1 was taken to the Clapham museum, or during the repainting work carried out after arriving at Crich. The mounting holes were still evident [89].

Since LCC No. 1's arrival at Crich, one of the cab doors had been removed and a display board placed at the entrance to the lower saloon, allowing visitors to board the tramcar but not to enter the saloon. A blanking panel had also been placed at the top of the stairs to prevent access to the top deck. These actions were considered to be reversible without causing any significant damage to the fabric of the tramcar [90].

As a result of the 2012 assessment, some short- and some long-term treatments were proposed to help stabilise the distortion. This would involve the removal of the saloon ceiling panels and possibly the lining panels, allowing inspection of the main carlines – the metal roof joints which follow the shape of the roof. This action was undesirable, however, as it would mean the removal of original interior features. The workload would be high and only a proportion of the steelwork could be removed. In the long term, the main frame would require work to slow the corrosion, including inserting a suitable barrier between the aluminium and steel components. This would also be invasive work, but it was seen to be inevitable intervention and would have to be undertaken in the future [91].

The paintwork on the lower deck ceiling had begun to fall away, revealing the ground beneath. Corrosion can also be seen along the metal window seal.
(D J Heeley)

A section of the flooring, showing the brown replacement covering put in place by Leeds City Transport to patch the original blue London County Council floor covering.
(D J Heeley)

One of the controllers; most of its paintwork has been lost, and the top has become heavily corroded. (D J Heeley)

The fuse board, showing its deteriorated wiring and missing parts as well as the corrosion to the metal surroundings. (D J Heeley)

Apart from those works carried out in the early years of its residence at Crich, LCC No. 1 still remained much as it had been when it left service in Leeds and was taken to London for display in the Museum of British Transport. If it had not come to Crich, it might have remained in storage, still in its Leeds livery, slowly degrading, its significance forgotten. Instead, partly by good fortune perhaps, but also largely due to the efforts of a number of people who appreciated its importance, it was destined to survive.

The story of its restoration – we might say, its rebirth – will be told in the next chapter.

Chapter 9
Restoration
Laura Waters

LCC No. 1 in the siding at Wakebridge, in London Transport livery, as part of the Museum's 2004 Enthusiast event. (D Yates)

An offer for restoration

The history of LCC No. 1 demonstrates that the path to preservation is not always an easy one, and that the tramcar owes its survival to the perseverance of those such as John H Price, who had the foresight and determination to see this unique vehicle become part of the National Collection.

In 2012, the Tramway Museum Society (TMS) were asked to consider further the offer of funding to restore LCC No. 1 to operational condition, thereby making it part of the Museum's demonstration fleet for visitors to experience.

This request once again came from the London County Council Tramways Trust (LCCTT), a registered Charity whose object as set out in the Trust Deed is "to educate the public in the history of tramways and the technical details, engineering and performance of such transport." [1]

The Trust had already funded the restoration of three other London tramcars which form part of the Museum's collection and were seeking a final decision on the restoration of LCC No. 1, having first offered to start fundraising for this in 2006.

As the National Tramway Museum (NTM) at Crich is an Accredited Museum with Nationally Designated collections, it was the responsibility of the TMS's Board of Management to carefully consider the offer and any implications of undertaking such a restoration. The Museum's Curator and Tramcar Conservation Committee (TCC) were asked to use their knowledge and expertise to advise the Board.

After discussion at its meeting on 10 November 2012 the TCC asked for a non-intrusive survey to be carried out "...in order to assist with decisions over the future conservation and/or restoration of the tramcar." [2]

The Museum uses condition surveys/assessments to assist in recording the known condition of a vehicle and to indicate areas of deterioration and concern, as well as possible future conservation or restoration options. Conducted by the Museum's Curator, Workshop Coordinator and a representative from the LCCTT, the survey report noted that the tramcar had "received little or no attention since withdrawal from passenger use in Leeds" and that failures in materials had led to water ingress and corrosion within the structure of the vehicle. It highlighted concerns that the corrosion, as well as electrocatalytic reactions between materials, was leading to distortion within the body, and although the vehicle was then stored in a fairly dry environment, the corrosion would continue without intrusive interventions.

Alongside the Condition Assessment, the Museum also utilises an Attitude Statement, a document which includes information relating to:

- Curatorial – Condition, Originality (body and equipment), Form and Function;
- Technical – Form, Function, Influence, Manufacturer;
- Social – Social representation;
- Health and Safety – Demonstration, Display;
- Financial – Restoration, Demonstration, Display, Maintenance;
- Other Value (commercial considerations) – Attraction, Operation, Workshop, Outside Works.

The Attitude Statement aims to establish the significance and position of an object in the Museum's collection, by presenting a clear and reasoned evaluation describing its history, value, meaning and importance.

The 2012 Attitude Statement proposed by the then Museum Curator for consideration by the Board of Management recommended that the tramcar be preserved as it was, and that a "...full Conservation Plan/Policy be drawn up to inform the future conservation and preservation of this important original vehicle." It further stated: "It is recommended that, after conservation work the car should be the subject of a dedicated exhibition that will inform the public of the significance of the vehicle and its design." [3]

A meeting of the Board of Management took place on 8 December 2012, during which they were presented with the results of the non-intrusive Condition Assessment and the proposed Attitude Statement from the Museum Curator, as well as the LCCTT's proposal, which included a review of the Attitude Statement's contents.

After considerable discussion and closer examination of the Attitude Statement's content and recommendations, the review submitted by the LCCTT, and the results of the non-intrusive Condition Assessment, the Board of Management on this occasion voted that LCC No. 1 "...be restored to an operational condition and that a full archaeological survey be conducted during the work... with a view to creating explanatory and interpretation media..." [4]

Funding

An initial estimate provided by the TMS's Conservation Workshop indicated at the time that the project to return LCC No. 1 to 1932 operational condition, including labour and materials, might cost in the region of £500,000.

For any restoration project to proceed, the Museum's Management ask that a sponsor or funding body must either commit or have available all the funds to complete the restoration, or demonstrate that they will have the funds in place when the project commences. In the 18 months between the Board of Management giving approval for LCC No. 1's restoration and the tramcar entering the Conservation Workshop, the LCCTT was able to raise the full estimated costs of restoration and commit fully to funding the project.

Preparations for restoration

"Before anything else, preparation is the key to success." (Alexander Graham Bell)

Due to the scale and nature of the project, many preparatory arrangements had to be made, and it would be 18 months - from December 2012 to June 2014 - before any restoration work could commence. The project team needed to develop and finalise the Restoration Specification, fully record the tramcar's existing condition, establish outline programmes and processes, draw up and sign funding contracts and terms, and, most importantly, identify a suitable space in the Conservation Workshop programme, once other projects had been completed.

Managing the project

Firstly, the project team, which would oversee and deliver the restoration, was assembled. It consisted of representatives from all the Conservation Workshop's different disciplines, together with members of the LCCTT, the project's sponsors.

Due to staffing changes, a new Curator would coordinate the Curatorial Department's input to the project. The Curator would be responsible for ensuring that the restoration met the requirements of the Restoration Specification, whilst the Conservation Workshop's Rolling Stock Engineer would be responsible for delivering the physical restoration work and delivering the project within budget.

The Restoration Specification

The Restoration Specification is the document the Museum uses to establish the standards and parameters guiding a project such as the restoration of LCC No. 1. Ultimately, it defines what the form and function of the tramcar should be when fully restored.

The document incorporates the Museum's overarching specifications for any restoration project. These state that the project should:

- meet the principles of the Museum's Mission Statement;
- be of such quality that the tramcar will have a minimum life of 30 years before further major conservation work is required;
- be safe to demonstrate on the Museum's tramway;
- meet any requirement of the Office of Rail and Road (the successors to H.M. Railway Inspectorate);
- meet the requirements of current Health & Safety legislation.

Further guidance in the document sets the criteria of how to meet the recommendations of the Attitude Statement, maintain historical accuracy and comply with the curatorial responsibilities of the Museum. The form and function of the tramcar must be accurately portrayed; this means that the tramcar must also be able to perform its original function, carrying passengers. The integrity of components – including service history and manufacture – must be preserved. As much of the original material as possible must be re-used, and what cannot be re-used must, if appropriate, also be preserved. Conservation skills must be maintained, and training provided if necessary; and any alterations to the tramcar made in order to fulfil the safety and interface criteria must be fully documented and reversible.

The third and final set of guidelines in the Restoration Specification deals with the challenge of restoring a historic vehicle, whilst balancing current Health & Safety regulations.

Any modifications made to the original equipment or structure are fully recorded and, if possible, undertaken in such a manner that the modified part can be reverted to its 'as found' condition, but there are certain cases in which modern materials and/or processes have to be used in order to meet current Health & Safety requirements. For example, any form of free asbestos would need to be removed and replaced with a suitable non-asbestos alternative, whilst all glass installed would need to be replaced with safety glass in accordance with British Standard BS857:1967.

In LCC No. 1's case the details for the Restoration Specification were focused on defining the characteristics of the tramcar during the period July 1932–July 1933. The revised post-December 2012 Attitude Statement had concluded that this was the point at which LCC No. 1 made its greatest contribution, not only to municipally operated tramways in London, but also to other tramway operators in the United Kingdom, with its design features being adopted in whole or in part by them [5].

Much of the research work undertaken for the development of the Attitude Statement, as well as a comprehensive timeline of the tramcar's history, pieced together alterations made to the vehicle throughout its operational lifetime. Further information had also been gathered through observation during the non-intrusive Condition Assessment. This research provided the framework for the details contained in the Restoration Specification, as it outlined much of the key 1932/33 characteristics of the tramcar, such as:

- high quality of internal and external style and livery;
- high standard of passenger accommodation including upholstered seating, improved ventilation, lighting and heating;
- improved accommodation for drivers including cab seat, windscreen and windscreen wipers;
- streamlined all-metal body construction;
- power operated platform doors and steps;
- trucks incorporating significant design improvements in suspension and braking systems.

With the Restoration Specification complete, attention could turn to other preparatory works, such as scheduling the tramcar's move into the Conservation Workshop and the programming of how the restoration would be undertaken.

Scheduling a move to the Conservation Workshop

The Conservation Workshop's programme is balanced primarily between major restoration projects, scheduled overhauls, and routine maintenance. Any delays to these works can affect the gaps available for the next major project.

During the period January 2013–June 2014 a number of projects were already in progress, including Sheffield 510 which was part way through a major overhaul. Scaffolding had been installed to assist with Sheffield 510's repaint, and this had an effect on any tramcars moving in and out of the Workshop.

As well as looking at the Workshop's programme, the Museum's Curator, Rolling Stock Engineer and Workshop Coordinator also considered the other tramcar moves needed to bring LCC No. 1 into the Workshop. For years No. 1 had been a part of the Great Exhibition Hall's various displays, and a replacement tramcar would have to be selected to fill the space which its removal would leave.

In 2014 a special scaffold platform was installed around Sheffield 510 whilst it underwent a major overhaul, including repaint, allowing the Museum's coach painters to access the upper deck. (National Tramway Museum)

During preparation for moving into the Exhibition Hall, conservation work revealed two advertisements on Edinburgh 35. These were stabilised by the Conservation Workshop's coach painter before the tramcar moved to its new display location. (National Tramway Museum)

Edinburgh 35 basking in the February sunshine, as part of the 'Big Shunt' event 2014, ahead of its move into the Exhibition Hall. (M Crabtree)

After much discussion and consideration of what would fit with the current Century of Trams timeline display, as well as what would fit physically, Edinburgh 35 was selected to replace LCC No. 1 in the Great Exhibition Hall. However, before it could take No. 1's place, Edinburgh 35 needed some conservation work to make it presentable for its new exhibition placement. This work needed to fit into yet another gap in the Workshop's programme, without delaying No. 1's arrival.

For members of the public and even the knowledgeable enthusiast, movement of the tramcars at the Museum can appear random at times. Often these movements are all part of the larger plan for the Workshop's programme of work, which has been carefully crafted to ensure that everything will be in the right place at the right time. In the case of LCC No. 1, the team at the Museum managed 18 months of meticulously choreographed moves, completion of projects, and temporary and permanent redisplays, all with the aim of releasing LCC No. 1 from static display and moving it to the Workshop for the start of its long-awaited restoration.

The restoration programme

Whilst the work described above was taking place, other members of the project team, led by the Project Manager, John Shawcross, were planning the restoration's programme of works.

When the Conservation Workshop provided their initial estimate for the cost of the project, they also provided a best estimate on how long they thought the work would take to complete: four years. During that time they hoped to completely deconstruct the tramcar to its basic framework and component parts, assess the deterioration of parts, work out how to reverse the changes made whilst the tramcar operated in Leeds, solve the dilemma of parts and materials no longer being available, restore, repair and renew all components and carefully put everything back together again. It was expected to be one of the most challenging restorations the Workshop had ever undertaken.

The team's project manager devised a top-level programme for the four-year period, which was then broken down with a sub-programme of work for each of the three main disciplines within the Workshop - Bodywork, Mechanical and Electrical. Within each of these were the outline tasks of what each discipline would need to deliver for the project, as far as was known at the time. However, when planning a project of this nature, the amount of work required is never known until the tramcar has been completely taken back to its basic structure and components.

Whilst carrying out research into the tramcar and the details of its form and function, the project team identified the following modifications which would need reversing as part of the project:

- bottom of side panels cut away to clear truck sides on sharp curves in Kirkstall Road Works;
- front and side service number boxes blanked out and side via point boxes reduced in size to accommodate standard Leeds destination blinds;
- bow collector fitted to roof and bow rope pulley wheels fitted on the roof ends;
- removal of plough gear;
- blanking off panel fitted between trucks to fill gap where plough gear had been removed;
- stair head drop windows replaced by sideways sliding type, also access handrail fitted to roof above this window to facilitate access to the bow collector;
- lower deck electric bell pushes removed and replaced by rope operated electric buzzers;
- curved glass window in front of platform doors replaced by steel panels;
- air-operated folding platform steps replaced by fixed steps;
- drop-down metal-framed windscreens replaced by new windscreens of an outward opening type with thicker wooden frames;
- remains of horizontal air operated windscreen wipers removed from top of windscreen frames;
- conventional windscreen wipers fitted;
- brake chains modified to prevent them from catching on the resistance grids;
- motorman's seats replaced;
- conduit power supply electrical equipment removed;
- compressor changed for a standard Leeds component.

Knowledge of these modifications gave an early indication of how the programme of works could be staged to allow the three main disciplines to coordinate, in order to achieve the works required to restore the component parts of the tramcar in the right sequence, then reconstruct it as a fully operational vehicle.

Long lead and missing items

Early on, as part of the programming work, the project team identified a number of items which were considered as possibly having a long lead time to source, and others which were missing entirely from the tramcar and would need sourcing or reproducing if originals could not be located. The items were divided between the different disciplines to group them together, and members of the project team were allocated tasks to start the process of sourcing materials or replacements. These steps were taken prior to any physical deconstruction work taking place, as some of the items were likely to be hard to source. For others, new parts needed to be manufactured, so the project team began further research in an attempt to find drawings and other known examples to use as patterns.

Labour resource

Anticipated to be one of the most challenging restorations the Conservation Workshop had ever undertaken, and to be completed within an ambitious estimated 4-year time span, the project would require a large body of resources.

As part of the funding agreement, the TMS and LCCTT had agreed to record not only the labour hours contributed to the project by the Museum's paid staff but also the volunteer input into the project.

The Conservation Workshop has across all its disciplines a range of paid skilled staff and knowledgeable volunteers, who contribute their time to assist with not only restoration projects such as LCC No. 1, but also the maintenance and upkeep of the tramcars which make up the Museum's demonstration fleet and are enjoyed by visitors to the Museum. Whilst primarily working within one of the three main disciplines, the Museum's staff and volunteers often cross over to another, as there are key intersection points where mechanical might affect bodywork, or where electrical intersects with mechanical equipment. The world-class restorations they deliver are a testament to the knowledge and skill of those who work and volunteer in the Museum's Conservation Workshop, and LCC No. 1 demonstrates the team's perseverance in tackling all the challenges and surprises that a tramcar of a bygone age can produce.

Saturday 14 June – the restoration begins

On Saturday 14 June 2014, after 18 months of preparation, tramcar moves, programming, research, and allocation of resources, London County Council No. 1 made its much-anticipated move into the Conservation Workshop at the National Tramway Museum, and the physical restoration of the tramcar to operational condition officially commenced. Now began the careful process of deconstructing No. 1, discovering its secrets, analysing its component parts and assessing their condition, and learning how to restore this unique 1932 experimental tramcar, so that passengers might experience again its luxurious comforts and experimental technology.

Saturday 14 June 2014 - a milestone moment as LCC No. 1 is shunted through the Depot gates on its way to the Conservation Workshop at the National Tramway Museum. (P Whiteley)

Restoration

In the following pages each of the main packages of work that made up the restoration programme will be explored, culminating in the completion of the project, the tramcar's commissioning, and its first demonstration in public service in over 60 years.

To assist in the recording of details during the restoration, the Museum's coach builder suggested using tablet computers to record the deconstruction process and component parts as they were removed from

A unique view: LCC No. 1 in the Conservation Workshop for the first time, alongside fellow London tramcars MET 331 and the remains of North Metropolitan horse tram 184. (P Whiteley)

the tramcar. The tablets would be portable enough to be easily used in and around the tramcar whilst work was taking place, also allowing the Workshop team to refer back to details quickly. This would ensure that fixtures and fittings were accurately placed as the work progressed from deconstruction to reconstruction.

The project team also obtained two shipping containers, providing more room to store parts which had been removed from the tramcar. In its deconstructed state, LCC No. 1 would occupy an area normally sufficient for three tramcars, not including the parts stored safely in these overflow containers.

Bodywork and associated mechanical work

1: Deconstruction

LCC No. 1, as described in the earlier chapters of this comprehensive history, is an all-enclosed double deck tramcar, with a steel underframe and steel main body frame to upper deck window level. Above this are the aluminium window frames below a single-piece plywood roof. Seating is fully upholstered on both upper and lower decks.

During the preparatory works for the project, an initial risk assessment highlighted the possibility of asbestos being present in several locations on the tramcar. Therefore, one of the first tasks for the team was to have the vehicle inspected by a specialist contractor. The contractor would take samples and test for the presence of asbestos; if any was present, it would be removed in a controlled manner. Twenty-nine different locations on the tramcar were assessed, and the presence of asbestos was confirmed in twelve of these.

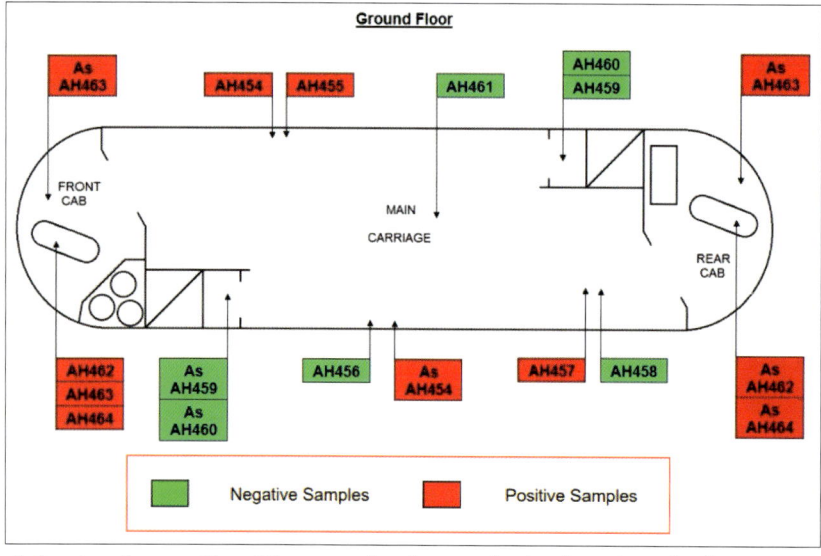

Asbestos Inspection Diagram for lower deck, showing the location of positive and negative asbestos sample results.

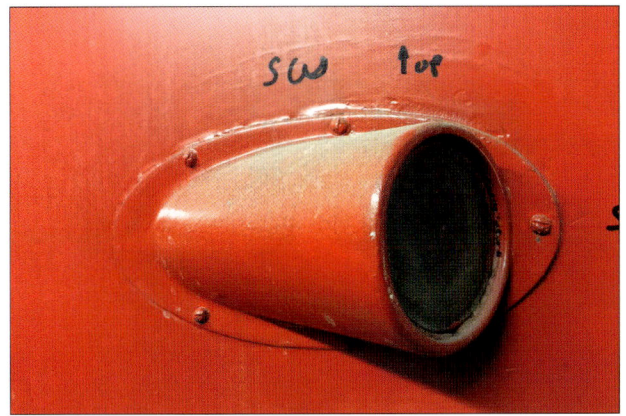

Small exterior fittings such as the torpedo-shaped 'Police lights' were amongst the first parts to be removed from No. 1, intriguingly revealing some of the first glimpses of the original blue livery. (National Tramway Museum)

As each seat base was removed it was given a unique location reference tag. The seats would all be returned to their original positions. (National Tramway Museum)

LCCTT Chairman, Ian Ross, taking a final look at the upper deck prior to the removal of fixtures. (National Tramway Museum)

Once the asbestos removal had been carried out, the deconstruction work could proceed. Starting in June 2014, surface-mounted items such as the exterior torpedo-shaped police lights on the upper deck and the door handles from the driver's cabs were among the first components to be removed and stored safely for assessment. The process continued with the removal of larger pieces of fitted equipment, such as the seating, windows, light fixtures, linings, air pipework and electrical wiring.

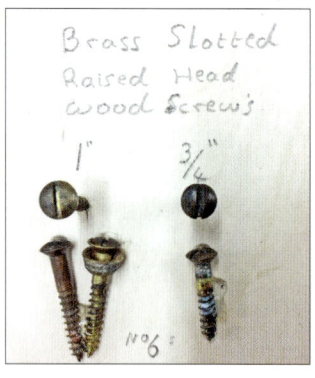
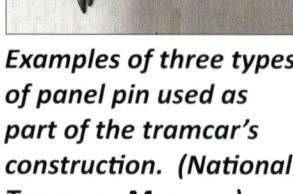

Examples of three types of panel pin used as part of the tramcar's construction. (National Tramway Museum)

As each item was removed, an initial examination was carried out by the team to establish the item's condition and to assess what future restoration treatment might be required, or if it would need replacing. Carefully peeling back the layers of fixtures and fittings allowed closer examination of the

Coach builder Richard Stead meticulously recorded details of all the fixtures and fittings, often creating measurement plans on spare bits of plywood, as here with the trims around the cove lights. (National Tramway Museum)

The lower deck ceiling panels, showing the size of each section and the wear and discolouration which had occurred since No. 1's days in service in Leeds. (National Tramway Museum)

changes made when the tramcar was in service in Leeds, and the Workshop team were able to establish the extent of the alterations as well as any decay and wear which had occurred as a result.

The lower saloon seat frames, which were bolted to the body sides and floor, were removed with spanners and sockets and where necessary with the aid of penetrating fluid, as, unsurprisingly, some joints had seized during the past 80 years.

The upper deck seat frames were discovered to be fixed differently and held with round head screws. It was more difficult to remove these as the slots tend to distort very easily when seized, so the task took the Museum's coach builder considerably longer than expected [6]. Each individual seat was given a reference location tag as it was removed, so they could be placed back in exactly the place they came from when they were refitted to the tramcar.

When the upper deck ceiling and the cove panels, which housed the lights, were removed, photographs of the large number of polished metal strips of different lengths, shapes and sizes were meticulously taken. The panels were also marked and labelled to ensure correct positioning after cleaning and replating. The lower deck light fittings were of a different design to those of the upper deck. The ceiling panels were fixed with panel pins; the ply panels were fixed to the roof sticks with pins of two different sizes, depending on whether they were fixed in the centre or at the edge. All panels, ceiling and linings were discovered to be of ¼ inch thick plywood.

One of the original luxuries of this 1932 experimental tramcar were the tube heaters, and whilst removing skirting boards the team discovered remnants of the wiring and asbestos lining for this original feature. The asbestos had of course been safely removed at the start of the physical works. It's thought that the heaters were probably removed sometime during the tramcar's service in Leeds, as there is no evidence to suggest they were removed whilst the tramcar was in service in London.

Screws found throughout the tramcar were of many different lengths, sizes, heads and materials. The lower deck lining panels, for example, used 1 inch or ¾ inch No. 6 brass slotted raised head screws.

The metal grilles hidden beneath the seats revealed the remains of the heater system, part of the original design and an added luxury for passengers in the 1930s. (National Tramway Museum)

During the deconstruction phase the team noticed hints that evidence of the original livery might lie beneath the top coat of red paint - then the original London County Council crest revealed itself in excellent condition. (M Crabtree)

These also have a flush screw cup. Not only was the position of screws recorded, but also the direction of slots. This attention to detail during the deconstruction period allowed the team to be incredibly accurate when the final restoration took place, as screw heads could be lined up to be exactly as they originally were.

Also during the deconstruction phase, attention was paid to identifying the original paint colours. Careful rubbing down through the layers of paint on a side panel, for instance, revealed the original London County Council crest.

A milestone moment occurred on 11 March 2016 when the plywood roof and upper deck aluminium window frames were removed as a single assembly from the main body frame. This crucial task required making sure absolutely every fixture had been undone and nothing still kept the structure attached to the body frame. In an operation which required the use of a large mobile crane, the roof and window frames assembly was transferred to a specially-made plywood platform mounted on accommodation bogies. Locating the roof assembly on the bogie-mounted platform meant that it would remain mobile and could be moved to different locations in the Workshop. This would allow other elements of the project to carry on, as well as ensuring that space remained flexible for the Workshop's routine servicing programme.

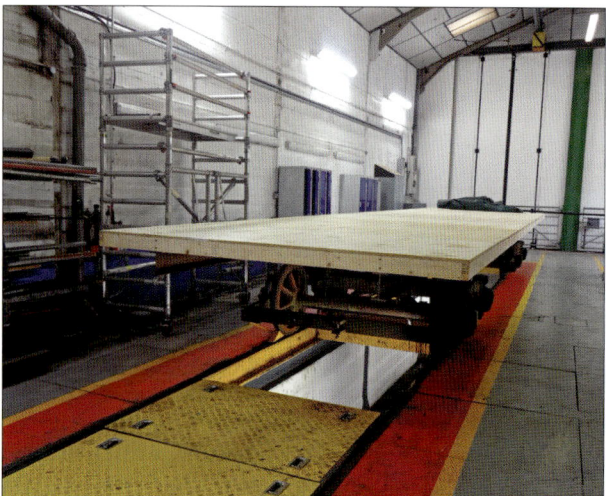

11 March 2016 - with assistance from Atilda Cranes, the Conservation Workshop completed the milestone task of removing the roof structure from the main body.
(M Crabtree/National Tramway Museum)

Finally, at the end of March 2016, work was completed on the removal of the platform steelwork for the tramcar, and the deconstruction of LCC No. 1's bodywork was complete. Just under two years since work had first started, the tramcar had been completely deconstructed back to its original basic steel body framework and upper deck single-length plywood floor boarding. It was fascinating to compare this unique sight to the contemporary images of No. 1's construction back in 1932.

A unique opportunity to compare original construction photos of LCC No. 1 (left) with the deconstructed form in 2016 (right). (LCCTT Collection/National Tramway Museum)

2: Analysis

During the deconstruction process, items relating to the tramcar's history and construction were uncovered, telling a more personal story of this unique tramcar affectionately nicknamed 'Bluebird', and allowing us glimpses into the lives of the original builders as well as the passengers.

The rubbing down of external paintwork by the Museum's coach painters to establish the number of layers and colours uncovered the remarkable and unexpected discovery of the original London County Council crest. The position of the original ivory 'streamlining' on both the upper and lower deck was also uncovered and its precise location recorded in detail.

 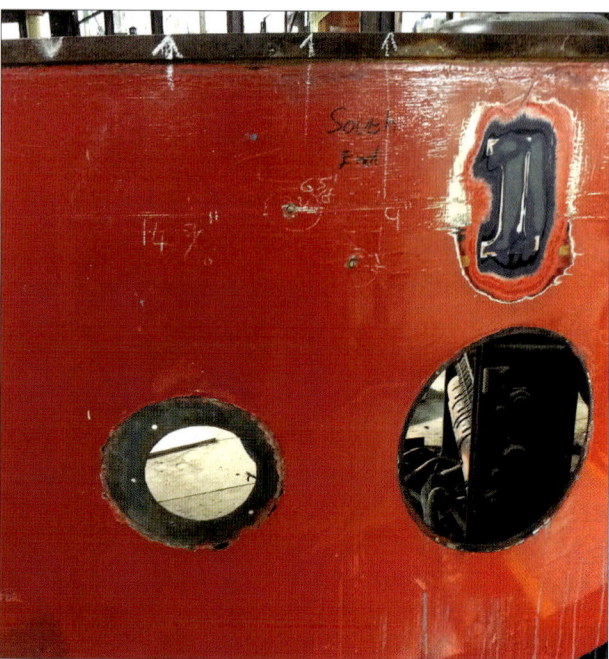

A key part of the deconstruction phase was looking for original livery details, such as fleet numbers and streamlining, that might be hidden underneath later layers of paint.
(National Tramway Museum)

The original London County Council crest survived in remarkably good condition under the layers of paint. (National Tramway Museum)

Various locations, including the north driver's cab's interior panel and the interior saloon panelling, moulding and skirting were all examined to help establish the differences in paint colours revealed during the deconstruction phase. (National Tramway Museum)

Initial examinations of the paintwork on both interior and exterior panelling suggested that potentially three different shades of dark blue paint were used on LCC No. 1. The team's experience from past projects and research suggested that this seemed an odd practice. Further investigation and a closer review of different panels concluded that the third shade, found on the interior dash panels, was in fact an undercoat rather than a finish coat.

Work on the trucks brought the Museum's coach painters into discussion with their mechanical colleagues. During their own deconstruction process, the mechanical engineers were discovering evidence of a mid-grey paint on the truck components and a ¾ inch wide blue painted line on the frame. They duly recorded as much detail as they could about these paint finishes for later reference.

The tramcar was giving up its long-held secrets, and for the Museum's Curator, each day of new discoveries disclosed a different part of the tramcar's history. Each layer revealed something new about how the vehicle had been constructed, and how different parts had interacted to create the form and function that was the experimental nature of LCC No. 1.

Surviving blue paint on the bogie frames. (M Crabtree)

Ticket boxes such as the one on the south platform of LCC No. 1 are prime locations for mementoes of the past to be discovered, including tickets used by former passengers. (National Tramway Museum)

Perhaps of equal, if not more profound interest for members of the Curatorial team were the curiosities that were discovered, which testified to LCC No. 1's life in service, and the people who operated and rode on the tramcar. The removal of the used ticket boxes from the lower deck, for example, revealed a selection of Leeds Corporation 'Ultimate' tickets; the boxes had never been emptied after the tramcar's last day of operation in passenger carrying service.

As the interior panelling beneath the drop windows in the lower saloon was removed, coach builder Richard Stead discovered a veritable timeline of tickets representing the history of passengers who had travelled on No. 1. (National Tramway Museum)

This child's ticket, issued in Leeds, is the pinkish ticket in the accompanying photo, pictured as it was found in the window cavity. (National Tramway Museum)

Objects such as combs, coins and post cards, pictured as discovered in the window cavities. (National Tramway Museum)

As the lower deck internal lining panels were removed, more tickets, in layers with varying quantities of dust and dirt, were uncovered in almost all locations. There were Leeds Corporation tickets with London Transport tickets below and at least one London County Council Tramways ticket below these. The London tickets were generally route specific for the combination 13, 33 and 35. A Scholar's prepaid London Transport ticket was also found.

Other, more curious items discovered among the accumulated debris behind the lining panels were a plastic comb, and a printed card from the Mayor of Islington, dated 1936. A Wills cigarette card depicting Stanford Robinson, conductor of the BBC Theatre Orchestra from 1932 to 1946, was also found. The card was one of a set of 50 Radio Celebrities and dates from 1935.

One or two items were specifically related to the conductor's duties, for example a London Transport Relief Way Bill Slip from Telford Avenue depot for 26 September 1938. This showed a midday working for LCC No. 1 on route 16, with nine values of tickets on the ticket rack from 1d to 8d for single, mid-day and return tickets.

From Leeds, one waybill dated 1952, others which were undated, and paper cash bags for coins were found. Some of these items were pushed into the fitting used to secure one end of the rope which acted as a barrier across the platform doorway, and others had been used to block the ventilation opening in the driver's door. Another relic of the tramcar's time in Leeds was a foreman's requisition note to the stores, requesting panel pins. This was found beneath one of the staircases. Was he repairing something on the tramcar, perhaps? It is fascinating to see these glimpses of the people who worked and rode on the tramcar during its operational life.

The Workshop team members, meanwhile, were equally fascinated by evidence of construction methods, together with modifications made as the tramcar was being built.

When the lower deck ceiling panels around the staircase were taken out, the team saw signs that a steel bracket had been readjusted to allow the prefabricated staircase unit to be accommodated; the original

Many of the items recovered from LCC No. 1 have now been cleaned and preserved in the Archive at the National Tramway Museum. (National Tramway Museum)

rivets had been burnt off at Charlton to enable the bracket to be repositioned. The same modification was found at both ends of the tramcar. The staircase housing in the lower saloon was found to be fixed by panel pins rather than screws.

Some of the interior trim panels had screw holes but no evidence of screws having actually been fixed. Also, some framing within the lower deck timberwork was only wedged in place without any fixings. Most timber and steelwork behind panelling was not painted, although some covered-in sections of the main pillars had been painted with red oxide paint of a type used in the 1930s [7].

These discoveries were a reminder that underneath the luxury finish of the tramcar with its stylish chrome fittings, comfortable upholstered seating and eye-catching paintwork were the bones of an experimental tramcar, a prototype, the first of its kind. The people who built it were still refining the actual construction and design as they worked. Finesse and refinement may have come had LCC No. 1 gone on to be the first of a mass-produced fleet of tramcars operating for London County Council in the 1930s.

There were, in addition, signs that other modifications had been necessary after LCC No. 1 had entered service, with physical details confirming evidence from historic photographs. The upper deck appears to have flexed soon after the tramcar entered service, causing fractures in the aluminium window frames. Around the seven windows at each of the upper saloon ends, the fractured aluminium had been repaired and strengthened with steel braces screwed to the inside of the framing to create a single element [8].

The end windows had also been modified, along with the destination boxes, soon after entering service. At each end the number box, originally made to display two characters, was modified to accommodate the width of three, and the single-piece end window was replaced by a vertically opening one.

London Passenger Transport Board Relief Way-Bill cleaned and preserved in the Archive at the National Tramway Museum. (National Tramway Museum)

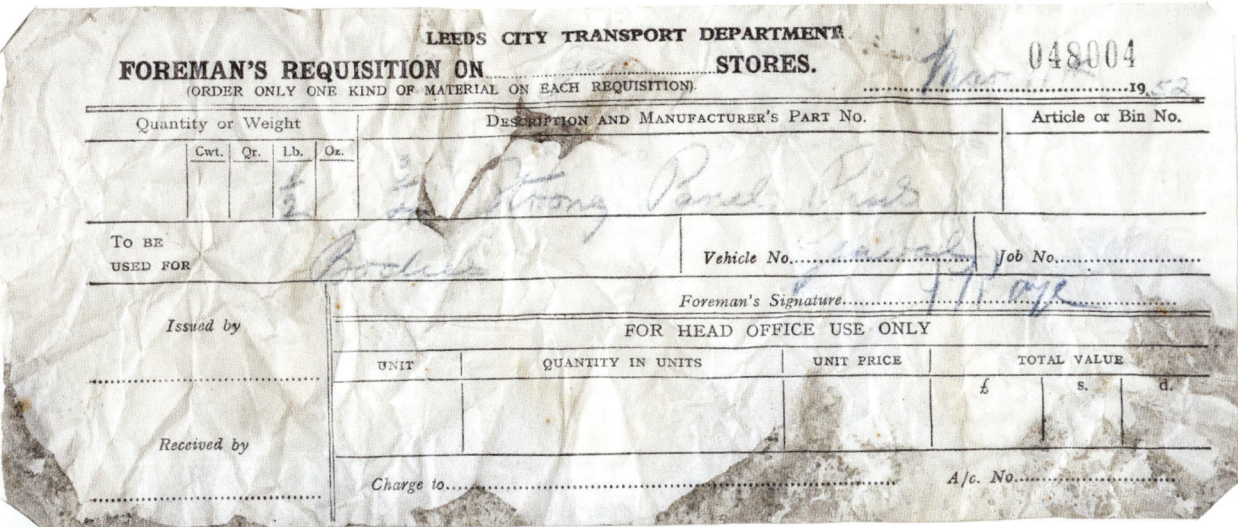

As each artefact was discovered it was bagged up carefully and the bags annotated with its location on the tramcar, for future reference. (National Tramway Museum)

The Conservation Workshop team found that some sections of the wooden structure beneath the interior panelling had been painted grey, some bore a red oxide undercoat, and others had not been painted at all.
(National Tramway Museum)

Various chalk and pencil marks were discovered when panelling was removed, a tangible link to the original coach builders who constructed LCC No. 1. (National Tramway Museum)

Evidence of the damage caused by water ingress in the window cavity of the lower deck drop windows. (National Tramway Museum)

A view of the exposed staircase structure. The rust stains at top right and the exposed holes mark the changes that were made to the positioning of the staircases. (National Tramway Museum)

A number of modifications were made whilst the tramcar operated in Leeds, including this hinged drivers seat. It would be replaced with a new original style pedestal seat made in the Conservation Workshop. (National Tramway Museum)

Significant corrosion in the panels and upper deck steelwork appears to have been due in part to the design and the construction methods available. The main side panels had some corrosion below the windows, particularly the four corner drop windows, due to poor sealing around the glass which had let in rain and tramcar washing water. The limited performance properties of sealants available to Charlton Works in the 1930s had resulted in water ingress below the windows on both decks [9].

Scorching on timbers below the staircases confirmed where electrical fires had occurred; in addition, when the electrical team recovered the wiring they discovered evidence of joints and replacement wiring. This had implications for the restoration work, as the tramcar's only wiring diagram was, according to Leeds sources, located here and had been destroyed in one such electrical fire.

All panels, ceilings and linings were of ¼ inch thick plywood. Some panels were larger than standard modern sheet sizes, so any present-day replacements would have to be sourced specially. The ceiling framing on both decks was in good condition, even retaining original chalked measurements for the panel sizes.

The hinged driver's seats, fixed to the doors in Leeds, had their origin in the 'Feltham' class of London tramcars, over 90 of which were supplied to Leeds second hand. This seat design was not original to LCC No. 1, therefore new seats would have to be made to fulfil the original seating provision for the driver.

3: Treatment

The condition of all the timber used on the tramcar was assessed as part of the deconstruction, and decisions were made on whether it could be reused in the restoration or needed to be replaced.

The Museum's coach builder identified several different types of timber used throughout the tramcar. Oak was used for mouldings, skirting and top deck light boxes, and ash for the roof sticks and framing of the lower saloon panelling. Light fittings and cove framings were made using softwood, and for other parts mahogany or mixed timbers had been used.

The lower deck flooring with new, lighter coloured wooden sections spliced in. (National Tramway Museum)

The hardwood of the flooring was generally found to be in sound condition. Splicing with new timber was necessary on some of the softwood framing and plywood sheeting used for the internal lining, where water ingress had occurred.

The interior fittings which had been carefully removed during the deconstruction phase of the project were taken from store, and the task of restoring and repairing progressed. After cleaning and inspection, the Museum's coach builders have skilfully retained more original material than was initially expected. Framing for the interior trim panels was repaired and refitted, and new leather cloth was applied on the panels.

Working steadily for nearly two years carefully recording and then removing all the component parts of the tramcar allowed the project team to see the state of the original steel frame bodywork.

The team had established that stripping the tramcar right back to the original framework was the best process to follow, in order to ensure the frame was structurally sound for the rest of the restoration to proceed. Doing so allowed the team to identify a number of elements within the framework which would require more specialist restoration work.

Because of the size and construction method of the body, it was necessary to sub-contract some of the work out in order to restore it. It was also necessary to identify a company who could make two new lower deck side panels. Each lower deck side panel is just over 21 feet in length and, unusually, each one is made from a single sheet of metal, as opposed to being cut into sections, as the upper deck panelling is.

New leather cloth was sourced to match the original material on the panels, which had degraded. Here, the process of re-covering the panels is shown. (National Tramway Museum)

The body as it arrived at Garmendale, with cutouts still present. (M Crabtree)

A number of visits were undertaken whilst the body was with Garmendale, allowing the team to check on progress and discuss techniques and restoration processes. (M Crabtree)

The contractor had to be able to make new lower deck side panels from a single sheet of steel, and to perform hot riveting to replicate the original method of securing the steel framework. Derbyshire engineering company Garmendale Engineering Limited were awarded the contract to work on this element of the restoration project, and in May 2016, LCC No. 1 was transported on a low loader the short distance to Garmendale's workshop in Ilkeston.

There, the 21 feet long original side panels were removed, with one section, which bore the original London County Council crest, being returned to the Museum. The side panels were removed and the frame braced to provide stability whilst working on it. The single-sheet side panels were restored; repairs were made where there was minor corrosion at floor level, and to derailment damage that had become evident in the floor supports. Some welds had not been completed when the tramcar was built at Charlton Works; these were remedied, as it was considered necessary in order to ensure future stability and safety of the body structure. Traditional hot riveting secured the panels onto the body frame.

The noticeable difference between the old panels and the new ones was the continuous straight lower edge along the whole length of the panel. This had returned the tramcar to its original appearance, as the cut-out sections on the old panels were a modification made when the tramcar operated in Leeds. The cutouts allowed the bogies under the tramcar to swing out clear of the bodywork on the tight curves in the Leeds overhaul workshops.

The whole of the top rail of the tramcar below the upper deck windows had to be replaced due to the corrosion of various sections. The original curved rail sections were used as templates for the new steelwork, in order to recreate the correct curvature.

Other tasks for the engineers at Garmendale Engineering included purpose building a tent to allow shot blasting of the stripped body. With 80-year-old paint on the tramcar, and because of the potential presence of lead in the paint, a very gentle shot blast method (by hand) was required. The team at Garmendale Engineering are experienced in this work and the result was a high standard of finish. Once the careful shot blasting and cleaning had been completed, the tramcar was given the first coats of primer, turning it a sandy red.

LCC No. 1 being loaded at Garmendale ready to return to the Museum's Conservation Workshop. The new continuous straight-edge side panels can be seen for the first time. (M Crabtree)

As Leeds had replaced the metal framing of the driver's windscreen with timber, the framework had to be redesigned for metalwork, and appropriate brass material had to be sourced [10]. One of the challenges the team experienced during the project came when fixtures and fittings had to be replicated from scratch. Not only did appropriate materials have to be obtained, but contractors with the skills to make such historic fittings had to be found. The windscreen framework was only reproduced after the team, following much searching, made contact with the specialist vintage car restoration company VBE Restorations, who were able to help.

The original tube heaters had been removed from the tramcar, presumably during its time in Leeds. Whilst evidence of wiring for them was found during the deconstruction, no further information was identified which would help to work out how they would have been installed and connected.

The Workshop team explored various options in their attempts to source an equivalent, and looked at other contemporary transport, such as the tube trains for London Transport; but without success. After much discussion within the project team and the wider Conservation Workshop team, the decision was made to reinstall the skirting as it had been found, leaving voids where the heaters would have been located.

The worn original curved sections of top deck rail were used as templates when new rails were made. (M Crabtree)

When undertaking a restoration such as LCC No. 1 there are times when compromises

 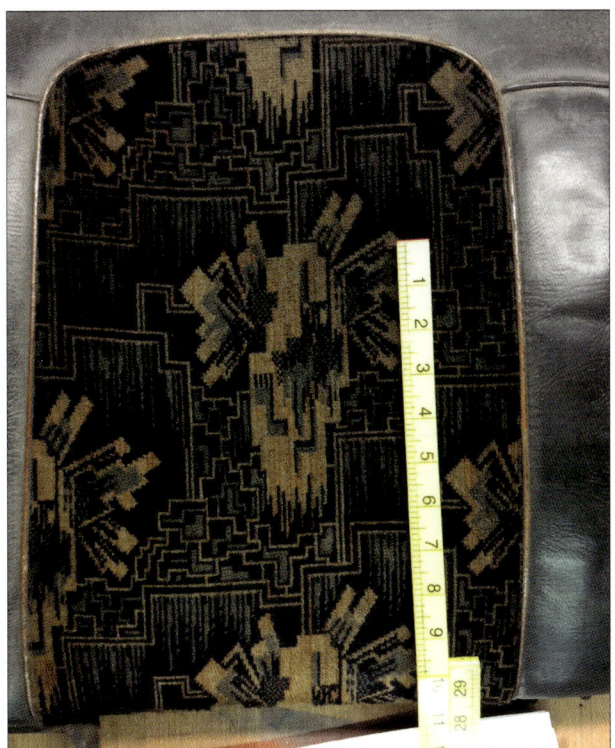

Close comparisons were made between a stock of existing moquette for tramcar LPTB 1622 (left) and the original moquette on LCC No. 1 (right) as the designs are very similar. (National Tramway Museum)

have to be made: if a feature cannot be restored because the relevant information is not available; or when fabricating and fitting something to replace the original would entail significant modification of the vehicle's original form. In this instance the team decided that leaving the heater locations empty, and retaining the details which had been discovered during deconstruction, was the best method, and would maintain their duty of care to restore the vehicle as accurately as possible.

Another fitting which proved a unique challenge to source was the roof-mounted ventilators, which are made of thin sheet metal. The only contractor available to carry out the work was found at Castle Caereinon in Wales. Similarly, a hardware store in Old Street, London was able to provide some suitable catches to replace much-worn originals. The team made great efforts to locate suppliers of original materials and fittings, in order to ensure accuracy.

For the lower saloon seating, the project team had originally planned to use an existing stock of moquette which the Conservation Workshop had in store. This moquette is carried by tramcar London Passenger Transport Board (LPTB) No. 1622, also in the Museum's collection. However, on closer examination of No. 1's original moquette, the Museum's Curator discovered that although the two were very similar in pattern and colouring, there were differences in the size and repetition of the pattern. LPTB 1622's moquette would not be right for LCC No. 1.

Camira Fabrics, the successors to Holdsworth, the company who supplied the original moquette for LCC No. 1's seats, were commissioned to supply new moquette fabric. Unfortunately, Camira couldn't locate the original design in their Archives, so they undertook the time-intensive task of redrawing the original design from a sample of fabric provided from one of the tramcar's seats. The process involved colour matching and producing test samples, which then had to be compared to the other seats on LCC No. 1 to check that the design had been correctly redrawn and the pattern accurately reproduced.

In the meantime, work continued on other components. Parts of the handbrake linkage which passed through the underframe were bent out of shape and required straightening in conjunction with some heat application. The platform doors were given new rubber edging seals, and the cab doors were modified to be fitted to their original opening direction, into the platform and not into the cab [11]. New steel channels and associated brackets were fabricated to replace the plough support and connection components which had been removed in Leeds.

4: Leeds modifications

To revert to the Restoration Specification: several major modifications made by Leeds, plus a number of minor modifications, had to be addressed.

Early tests were carried out by towing LCC No. 1 around the Museum's Depot fan, to check that it would be possible to restore the full side panels and that the tramcar would then operate on the Museum's track. (National Tramway Museum)

Prior to the completion of its deconstruction, LCC No. 1 was towed over all the trackwork on the Museum's Depot fan, to establish whether reinstating the lower deck side panels to their original design and reversing the Leeds modification of cutouts along the bottom edge would leave sufficient clearance for the swing of the bogies. The clearances were found to be adequate.

Two upper deck windows had been changed in Leeds from the original drop design to a sliding variation. The sliding windows allowed access to the roof and bow collector, and also allowed the conductor, when upstairs, to see passengers on the back platform.

Whilst restoring the missing drop windows on the upper deck, the Conservation Workshop explored new techniques including X-Ray to try to understand how they were constructed, as no original drawings were available. (National Tramway Museum)

This modification was one of the more challenging ones for the project team to source materials for, as the original window rack, catches and trim had been removed when Leeds introduced the sliding window. No. 1 still had examples of the original drop style window, and the team were able to look at these to identify the component parts. One of the original windows was sent away to be X-rayed to see if it would reveal more details about how the rack and catches operated and what parts were needed to reproduce them. However, even with this investigation the team were finding it hard to source the required brass section to the original specification.

Fortunately, an offer of help came from a TMS member who was also part of the Grimsby and Cleethorpes Model Engineering Society, and who thought they might be able to assist with the complexities of making two new drop windows.

After much investigation work and more examinations of one of the existing windows, a compromise in the use of 'U' shaped brass channel was reached, as it was impossible for either the Conservation Workshop or the Grimsby and Cleethorpes Model Engineering Society to source this material. Instead, it was agreed that an alternative method of fabricating this particular element of the window design would be used. The result would be virtually indistinguishable from the original, but the process would be fully recorded for future reference.

Members of the Grimsby and Cleethorpes Model Engineering Society investigating the construction of one of the original drop windows. (G Marsden)

In March 2019 the curved glass was restored to the corner structure of the platforms. (National Tramway Museum)

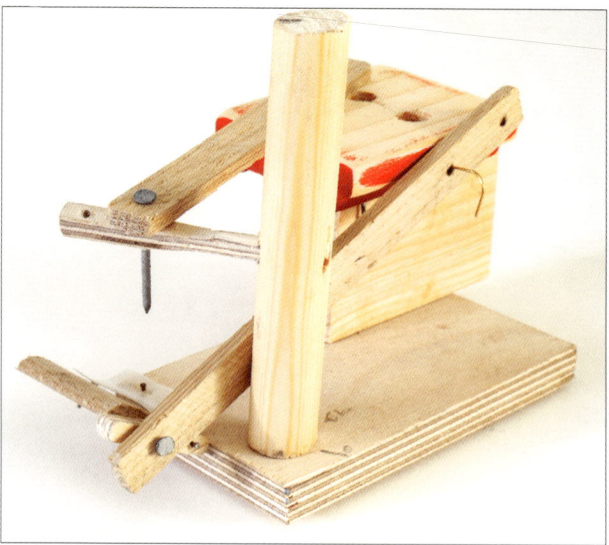

Initial wooden maquette of the folding step mechanism built by Workshop volunteer Richard Sykes to understand how the original step would have functioned. (National Tramway Museum)

Finally, the two missing drop windows were restored to the upper deck, and the curved platform glass on the cab ends, which had been replaced with metal panels, was reinstated.

The destination boxes were modified by Leeds for their form of destination layout. The original specification has been achieved by appropriate woodwork and installation of gearing by the Workshop's coach builder.

The hinged driver's seat was removed from the door in each cab, and a fixed pedestal seat produced.

One of the unique features of LCC No. 1 was its folding step mechanism which was integrated with and driven by the platform door opening mechanism. It included the lifeguard which would swing into place when the step was closed. The folding step had been removed in Leeds and replaced with a fixed step, and whilst the door engine and door fittings remained in situ, a lot of reverse engineering was required for the folding step mechanism to once again function on the tramcar.

The team spent nearly two years, between November 2017 and November 2019, researching and exploring how the folding step mechanism could have operated. Again, the team lacked technical drawings, but they considered evidence found on the tramcar, studied contemporary photos and film footage, and examined other similar preserved steps, such as that on the Museum's New York 674 tramcar, in an attempt to understand how all the parts interacted and to decide what would be needed to restore the mechanism to working order.

Following more research, a basic prototype was made by the Conservation Workshop's mechanical engineers. (National Tramway Museum)

The Conservation Workshop built several prototypes, from small wooden maquettes through to full-size replicas, before temporarily trialling a prototype they thought was close to the original design and function.

In March 2022, Workshop mechanical engineers

The completed folding steps on LCC No. 1 in the lowered position (left) and the folded up position (right), including the curved covers now concealing the working mechanisms.
(National Tramway Museum)

Neil Rowland and Andy Parry completed the design, fabrication and installation of a working step mechanism that is as close as possible to the original concept installed by London County Council Tramways in 1932. This reverse engineering of a complex mechanism has been achieved through hours of hard work, and is a testament both to their skills and their commitment to fulfilling the restoration specification.

New trolley bases utilising some components from Blackpool, trolley poles and trolley wheels had either to be made from available material or machined from new, as these items were all replaced in Leeds by a Fischer bow collector.

5: Reassembly

Since the only original official drawing of LCC No. 1 was a general arrangement, the reassembly was a direct reversal of the deconstruction, except where changes to meet the Restoration Specification were required.

While the main body frame was being modified, one of the first restoration tasks was to re-canvas the tramcar's roof.

Re-canvassing LCC No. 1's roof
In preparation for the re-canvassing, the roof had been stripped of all the old canvas, rubber lining and glue. The roof was then painted, and a new rubber lining added to the curved front end sections.
Previously, tramcar roofs at the Museum had been re-canvassed in sections, since pieces of canvas large enough to fully cover the roof had not been available. On this occasion, canvas material was sourced which was both wide and long enough to enable the roof to be canvassed in one piece, as originally seen on the tramcar.
On the day prior to re-canvassing, glue consisting of a butyl-based sealing compound thinned down to brushing consistency was prepared.
The operation was begun by placing the canvas at the centre of the roof; it was to be glued down in stages, working in sections across one half of the roof, before returning to the centre and completing the other half.
Once the glue had been applied to the first section of the roof, the canvas was pulled tight and smoothed over with what was essentially a flat iron, to remove any wrinkles in the canvas or bubbles of glue. At the same time the sides were pulled down tight and fixed with small pins along the edge to hold the canvas tightly in place. This process was then repeated until the entire roof was canvassed.
The first coat of paint was then applied to the canvas. This layer was a diluted version of the final roof paint. Applied directly after the canvas had been glued down, it acted almost like a second adhesive to secure the canvas more firmly to the roof. The paint dries quite quickly. The coach painters then continued to build up the layers of paint on the canvas to create a watertight finish.
Timber mouldings were applied to trap the edge of the canvas beneath the roof, making it watertight around the edges, and bearers for the trolley planks were fitted to the roof.

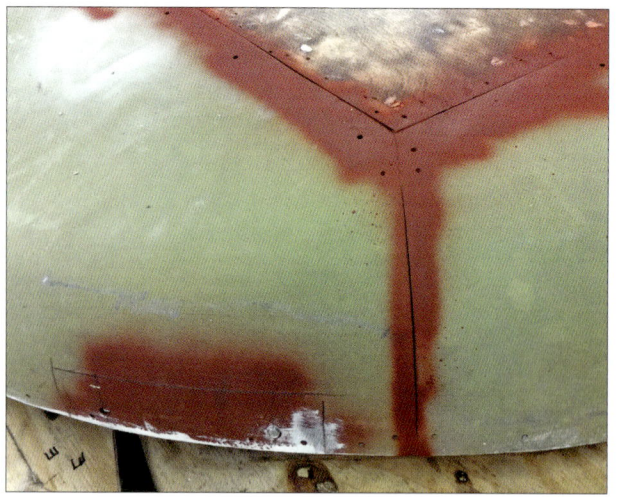

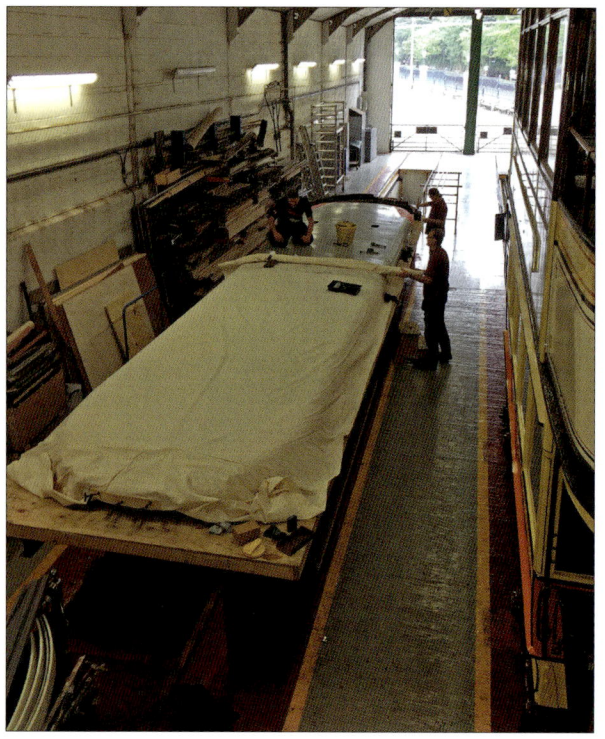

After the restored body frame was returned from Garmendale Engineering in June 2016, work started on rebuilding the platforms.

The upper deck window frames were dry fitted with the re-canvassed and restored roof, and once the team was happy with how these components had been reunited, the roof assembly was ready to be refitted to the body frame.

In November 2016 the large crane was once again hired to assist with the operation of reuniting the roof assembly and body frame. With the Museum closed to the public for the winter it was the ideal time to undertake the operation, which required the temporary movement of a section of the overhead wiring. Since the roof section was not built to be a stable structure in its own right, temporary support had been provided on a specially constructed mobile platform. This platform was towed to the waiting crane.

A system of spreaders, designed to support the load evenly, were attached and the complete roof assembly was lifted carefully from the platform.

The steel-bodied framework of the tramcar was positioned beneath the roof assembly which was then lowered, aligned and fixed to the main body. The complete structure was returned to the Conservation Workshop, where the main body continued to be built up with both platform structures in place and lower deck flooring timber installed. New upper deck steel side panels were fitted [12] [13].

The interior trim facings of Rexine had been removed, but the actual panels were only renewed where the plywood had been water damaged. The rest of the original panels were retained and restored. The panels were covered with a modern leathercloth version of Rexine, matching the original dark blue shade. The beading and skirting trim had been retained as far as possible; it was stripped of old paint, sanded down and repainted before fitting.

The upper deck flooring received a layer of plywood to provide a base for the linoleum sheet finish. In early 2019, glass was fitted to most of the windows, the ceiling panels were reinstalled in the upper saloon and the first of the newly chromed fixtures and fittings were added to the tramcar.

Upper deck window frames being dry fitted to the roof. Dry fitting is the process of checking everything is aligned correctly before committing completely to the final fitting process. (National Tramway Museum)

Newly restored platforms being rebuilt, with original components being installed and freshly painted for future preservation. (National Tramway Museum)

In early 2019, the first glimpses of the restored luxury interior of LCC No. 1 started to appear as new linoleum was laid on the upper deck and refurbished chrome fittings were restored to their original locations. (National Tramway Museum)

On a slightly snowy December day, a now blue No. 1 was moved to another location in the Conservation Workshop. (National Tramway Museum)

After much sourcing and comparison of colour swatches, the team identified a supplier of blue linoleum, and this was fitted to the upper deck floor. No. 1 was once again starting to regain its luxurious 1932 design, and its glamour.

The front destination blinds and service number blinds were installed. These had been specially made by Roy Makewell of the East Anglia Transport Museum. His research and study of photographs of LCC No. 1 taken during its early years in London ensured that the new blinds were as nearly identical to the originals as possible.

Using the photos taken during deconstruction and the location tags which had been applied during that process the restored seat bases were refitted to the upper deck in exactly the location they had been removed from.

With Camira having successfully reproduced the original design of the moquette, Nottinghamshire transportation seating specialists, Grinsty Rail, were contracted to re-upholster the seat cushions for both the upper and lower decks. The project team reached another milestone moment on 29 October 2019 when the first freshly re-upholstered seat base was installed on the upper deck.

In late 2022, the completion of the lower saloon also started to come to fruition. As the traction and the lighting wiring, which was to be routed

The first trial fit of the front destination blinds: a reminder of routes travelled when the tramcar was originally in service. (National Tramway Museum)

Newly restored fixtures such as seat bases were fitted more rapidly on the upper deck than on the lower. More work on the wiring was required on the lower deck, delaying the fitting there.
(National Tramway Museum)

With additional fittings and final finishes still to be completed, the newly installed seats and lino flooring on the upper deck were covered to keep them in good condition during the final stages of fit out. (P Bird)

October 2019 - the first fit of the freshly-restored upper deck seat pads restored comfort to LCC No. 1's upper deck.
(National Tramway Museum)

Conservation Workshop coach painter Matt Linaker spent hours painstakingly building up the layers of paint on the restored body panels, working through the process of applying paint, flatting it off, rubbing down and building up again, as he worked towards a smooth surface and a final mirror-like finish.
(M Crabtree)

The lower saloon ceiling panels partially refitted, as final electrical tests are carried out on the wiring within the ceiling voids. (R Sykes)

A view of the lower saloon progressing toward final fit out. (National Tramway Museum)

through the lower saloon ceiling voids, had to be installed first, much of the work to refit the saloon came late in the restoration programme.

As the wiring was tested and proved to function successfully, the lower saloon ceiling panels could be refitted, the flush mounted lighting fixtures installed, and the final coats of paint applied.

With no further paint or installation work required to the ceiling, attention could be turned to preparing the lower saloon's floor, ready for the installation of its lino covering and the perforated grilles for the saloon heaters. As mentioned earlier, despite hours of research the grilles are the only indicator of the original heater system for LCC No. 1, as replacement parts could not be sourced or replicated.

Finally, as with the upper saloon, the newly upholstered seating was returned to its previous locations, and passenger comfort was once again a major feature in the lower saloon of the tramcar.

One of the last tasks was the installation of newly fabricated plough carrying components which are attached to the underside of the tramcar.

As with the upper saloon, the lower saloon lino flooring was replaced with new. (R Sykes)

LCCTT Chairman Ian Ross stands as Workshop coach builders Richard Stead (left) and Brian Bates (back right), along with Workshop volunteer Bob Blackwell (front right), enjoy the re-installed luxury seating of LCC No. 1's lower saloon. (National Tramway Museum)

From the beginning of 2018 both exterior and interior paintwork was progressed. A great deal of filling and flatting down was carried out on the lower deck sides and dashes to take on the first coats of paint to achieve the blue and ivory livery [14].

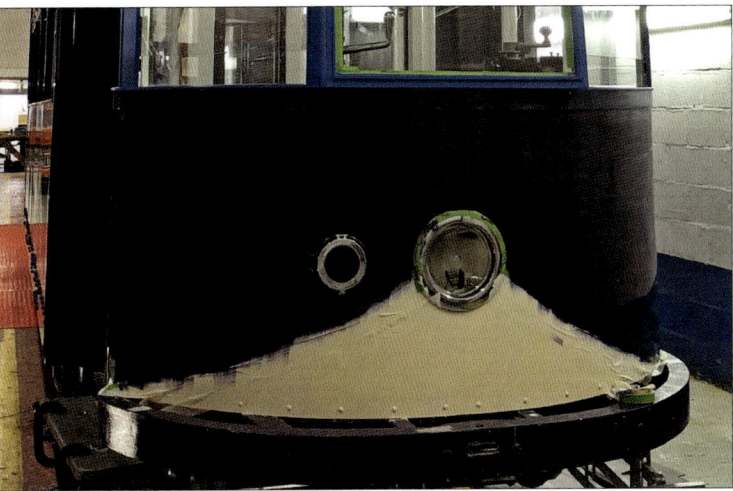

A close-up view of work ongoing to re-install and paint the interior trim of LCC No. 1.
(National Tramway Museum)

This view of the front dash shows the masked-out area for the ivory streamlining being given another coat of paint to build up the layers to final finish.
(National Tramway Museum)

The appearance of the ivory streamlining on the dashes was achieved by string and pencil with the centre being at the top edge of the windscreen [15]. The upper deck streamlining was applied in late 2019.

In May 2022 No. 1's appearance changed yet again, bringing the tramcar another step towards its finished state, as professional signwriter Trevor Kervick joined Matt Linaker, the Workshop's coach painter, to work on the adverts that were to be restored to No. 1.

The striking adverts break up the plain blue of No. 1's livery but are also a reminder of the attitudes of the 1930s when people were actively encouraged to enjoy not only the comfortable tramcar ride, but a smoke as well.

In February 2023, some of the tramcar's final livery details were being completed, including the replication of the ¾ inch lining on the bogies, discovered by the Workshop's mechanical engineers during the initial stripping down.

A blue-sky testing day on the Museum's Depot fan afforded a clear view of the first of the newly restored adverts on No. 1. (I Ross)

157

Professional signwriter Trevor Kervick making a start on the adverts to adorn LCC No. 1. (I Ross)

The second advert now restored to LCC No. 1 advertises the fares of the day for riding the car in the 1930s. (National Tramway Museum)

As the restoration headed into its last months, a final livery detail was settled, with the clarification of the correct colouring of the crest that LCC No. 1 carries.

The crest discovered under the layers of paint in the early months of the project had discoloured over time due to the darkening and yellowing effects of the varnish. Particularly in need of solution was the question of the colour of the waves, which are officially described as "barry waves of six azure and argent". 'Azure' means blue, but the correct shade had still to be decided upon.

The team did consider whether the blue on the discovered crest had discoloured or whether that was the correct shade of azure, but on the greener side. However, the crest is of a standard LCC design, and further research using the photographic sources from the Heraldic Society, a film called Capital County (1951) from the London Metropolitan Archive, the model of LCC No. 1 held by London Transport Museum, and sources at the London Ambulance Service Historic Collection, confirmed that the correct shade was a much darker azure blue, as can be seen in the newly recreated artwork for the transfers that now adorn the tramcar.

Close-up view of the blue lining starting to appear on No. 1's grey trucks. (L Waters)

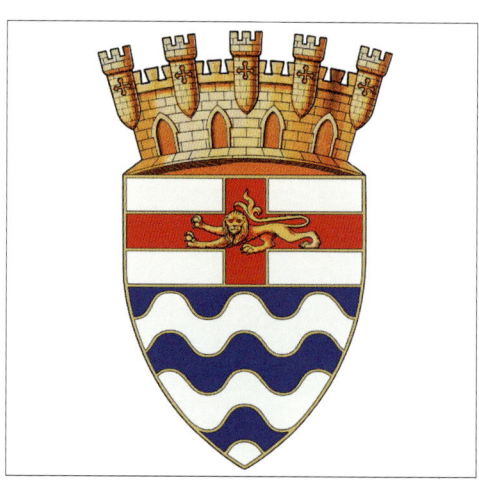

The digital artwork for the recreated and correctly coloured crest, that now adorns LCC No. 1. (Produced by Alan Padfield, Chapel Graphics)

Trucks and associated electrical work

1: Deconstruction

LCC No. 1 is fitted with a pair of EMB Class 6A equal wheel bogies, each with two traction motors. Once the tramcar had been moved into the workshop, the project team made the decision to delay the start of work on the bogies. In this way, the tramcar would remain mobile, enabling the team to move it to different locations in the Workshop for as long as possible.

May 2015 - the body is separated from its running gear. A team effort ensured that this was done smoothly and successfully. (I Ross)

In May 2015, almost a year after the tramcar had moved into the Workshop, it was finally raised on lifting jacks and the body was separated from its bogies. The bogies were sent to a local industrial cleaning firm for pressure washing, to remove the dirt accumulated during years in preservation. The Workshop environment would then be kept clean whilst the bogies were dismantled [16] [17] [18].

The motors and wheelsets were removed from both bogies, along with the track brakes and leaf springs [19]. It was decided that due to the complexity of the bogies, one would be carefully dismantled before the other to avoid possible mixing of components. This was also necessary because of the limitations of space to store the removed components and assemblies.

London Transport Museum's HR/2 truck, on loan to the Museum during the project, on display in the Great Exhibition Hall. (M Crabtree)

To aid the process of overhauling and analysing these complex bogies the Museum was grateful for the London Transport Museum's loan of its London Class HR/2 tramcar bogie held in its collection at Acton. Combined with a number of Electro-Mechanical Brake Company Ltd drawings held by the Museum's Library & Archive, this bogie would act as a 3-D model to check work against and make sure the team understood how the components of the bogies worked together and connected.

A first look in daylight at the condition of the removed bogies before restoration work started. (I Ross)

The bolster casting, which is the main crossmember for the bogie, has a complex secondary suspension consisting of a horizontal coil spring system, and, beneath it, a linking system of control rods. It was necessary to turn the truck frame over so that work could proceed satisfactorily on the bolster castings and their associated components. They were then removed, leaving the channel section transoms, which were in a poor condition due to distortion.

159

One of the original casting marks which still remained on LCC No. 1's tyres. (M Crabtree)

Machine work being done in the Conservation Workshop by the Museum's mechanical engineers and volunteers. (M Crabtree)

2: Analysis

As the bogies were dismantled, it soon became evident that restoring them to working condition was going to be one of the most complicated pieces of work the Museum's mechanical engineers had undertaken in the Conservation Workshop for some time.

Most of the brake linkages had holes for the connecting pins bushed to keep the tramcar operational, along with badly worn pins. Many holes, even with bushes, were found to be badly worn, indicating that the tramcar had had no major overhauls during its latter days in service. These bushes were probably fitted in Leeds, since London would not have replaced worn linkages prior to dispatch to Leeds.

New metal bushes were fitted to repair this part of the bogie assembly. (M Crabtree)

Bushing
Various rods and brackets form the brake linkages and are connected together by large pins. During the operational life of a tramcar, the holes for the pins will gradually wear until eventually the pins are allowed excessive movement. At this point either new rods or brackets have to be made, or the holes must be repaired by bushing.
Bushing on LCC No. 1 has been a repair method used to avoid waste by prolonging the life of components which wear in daily use.
For the bushing, the holes are drilled to an enlarged diameter so that they are circular again. The holes are now too large for the pins. A new ring of metal, the bush, is pressed into the newly enlarged hole in the rod or bracket. The external diameter of the bush must be such as to make it a force fit to the new hole, whilst the internal diameter must be the same as the original pin size.

The motor suspension brackets mounted on the transoms were badly worn. These were secured by set bolts and the heads of the bolts were found to be welded to the transom. This was probably done to stop them loosening, and initially the Museum's mechanical

LCC No. 1's wheel sets and motors, in the foreground, await transport to specialist contractor Dorlec Ltd for investigation and overhaul. (M Crabtree)

engineers thought this to have been a Leeds solution. However, the same welding was found on the London Transport Museum HR/2 truck suggesting it was a fitters' practice from the 1930s, rather than a designed solution, for a more commonplace problem in London.

The channel section transoms, as mentioned above, were in poor condition, showing distortion in shape; one was also visibly cracked [20]. The dimensions of the transoms were compared with the original drawings held in the Museum's Library & Archive, which indicated the channels fitted were of a lesser specification, being of a lighter section.

Replacement channel matching the drawing specification was not available commercially in 2019 and the original specification material may

Original marking discovered on one of the axles. (M Crabtree)

not have been available in 1932 either; so the nearest alternative may have been used for LCC No. 1, thereby demonstrating a contemporary need to find an alternative material source to be able to continue with the development of the tramcar.

Investigating the bogies during their deconstruction was confirming suspicions that limited regular overhaul work had been carried out on them. The markings on the wheelsets were another interesting discovery. Some of the wheel centres were dated 1932, whereas some of the tyres had the imprint '1952', indicating retyring by Leeds.

During the process of removing paint layers from the bogies, the team made an unexpected discovery: the original colour had been mid grey, with a blue lining around the frame edges.

3: Treatment

The wear on some bushes on the brake rigging and suspension was so great that they had worn through to parent metal. The treatment necessary was to further mill out and rebush with larger bushes. Although the team were working on one bogie at a time, a production line for new parts was introduced. In this manner over 100 bushes were produced to cover the two bogies' needs, as a similar level of wear was expected to be found on the second bogie.

As soon as an item was repaired it was labelled and put into stock before removing and repairing the next. All the components were cleaned and then painted with primers, undercoat and gloss finish.

The brake rigging was dismantled and the swing links removed. A number of new elements replacing worn items had to be fabricated.

The four motors and wheelsets removed from both bogies were sent to Dorlec Limited, a local Derbyshire electro-mechanical engineering company [21]. The motors went through a routine repair process; the armature shafts passed non-destructive testing (NDT), and were tested for insulation and varnished, while the commutators were skimmed in a lathe and the bearings replaced.

Previously unidentified problems which could not be identified by external examinations were found within the motor carcases, with field coils in need of rewinding or reinsulating. The overhaul included new suspension bearing shells and bushing for the motor suspension bolt holes [22].

When the wheelsets were non-destructively tested, two axles were found to have flaws. As there was wear in many of the bearings and on at least two gear wheels, new gears and pinions were made and all four axles renewed. New roller bearings for the axleboxes were also obtained [23].

It was clear that the flawed axles must be either repaired or replaced, or serious problems could arise when the vehicle went into operation.

The team had to consider carefully the retention and renewal of original materials versus replacement with new, in order to meet the remit of the Restoration Specification. This situation can – and usually does - arise during any restoration project, and the Curatorial and Engineering teams must always be prepared to consider such compromises; it is all part of the challenge of restoring a historic vehicle to an operational condition.

Finally the decision was taken to renew the axles, which would entail replacing the worn and flawed components with a significant amount of new material. The next challenge for the team was how to preserve the historic information from the originals, as there would be an irreversible loss of material in the process of changing the axles.

A photographic record of markings and stamps was crucial. In addition, measurements and particulars of the originals would be recorded, and where possible some of the original material would be preserved for future display or research.

The process of reuniting the gears and wheels with the new axles. (M Crabtree)

For a tramcar to operate on the Museum's track, the maximum distance from the back of one wheel to the back of the other wheel on the same axle must be between 1,390 mm and 1,392 mm. When LCC No. 1 was operating in Leeds, its trucks were retyred with a back to back measurement of 1,398 mm, too wide for Museum operation. A novel method was tried on the nearly new tyres fitted by Leeds; selective heating was used to expand each one and separate it from the wheel centre. To reduce the back to back, 5 mm was machined out from within the tyre depth, reducing the lip at the outside of the tyre. The tyres were then refitted and machined to the Museum's wheel profile.

The rolling road test: another milestone moment, probably the first time No. 1's motors had run since 1957. (M Crabtree)

The bearings during a pause in the rolling road test. (M Crabtree)

New roller bearings for the axle boxes were fitted and bushes on the radial arms were renewed.

Dorlec Limited, who had been working on the motors, also reassembled the wheel sets. The reassembly entailed freezing the axles and heating the gears and wheels, and then combining them together in a narrow window of time.

This process had to be repeated on each axle in order to get both wheels in position, along with the gear. The four wheelsets (eight wheels and four gears) were completed the same day. After painting, the wheelsets were reassembled with the overhauled motors.

Reuniting Axles and Gears
One of the axles is dipped in liquid nitrogen to cool the metal, which then contracts, reducing the diameter of the axle. The wheel centre and gear are gently warmed in an oven, which has the opposite effect of the liquid nitrogen, and allows the metal to expand.
The wheel and gear are removed from the oven and positioned exactly where they will be in relation to one another on the axle. They are placed the correct distance apart using a removable distance spacer. The axle is then lifted out of the liquid nitrogen and dropped through the gear and wheel. An end stop underneath the wheel prevents the axle dropping through.
Once the axle has been dropped through the gear and wheel, the assembly is allowed to return to the ambient temperature of the room. As it does so, the metal of the axle expands and the gear and wheel contract, creating a firm bond between the two components.

Before they were returned to the Museum, the motors and wheelsets were tested as part of the overhaul. A complete wheelset and motor were found to weigh 0.9 tonnes [24]. Two of the four motors and wheelsets were tested initially, using a rolling road which had been jointly designed and developed by the Museum's mechanical engineers and the engineers at Dorlec Limited. The rolling road test gives a check on the performance of the commutator, gear interface and bearings. This can be difficult to do once the truck has been reassembled and placed back under the tramcar's body.

The performance of the motor was monitored during the test; this included checking that the bearings were running at the correct temperature. The test was a significant point in the restoration of No. 1's trucks, as it was likely that this was the first time the two motors had run since 1957.

The other two motors and wheelsets were also tested. Once testing was complete, two of the motor wheelsets were returned to the Museum to be reunited with the first of the two bogies, and Dorlec stored the other two until the Museum was ready to receive them back for work on the second bogie.

The transoms, which were distorted as described above, had been remade to the correct specification from solid steel billets by CMS-Cepcor at Coalville in Leicestershire.

To establish some packing details at the bogie and body bearing points the underside of London Transport 1858 at Carlton Colville Museum was examined as a comparator with similar attributes. A spacer, previously thought to have been inserted by Leeds, was established as being an original fitting [25].

4: Leeds modifications

While at Leeds, the trucks had received both major and minor overhauls, but no modifications, remaining basically unchanged since their time in service in London. The only change attributable to Leeds was that of paint colour, from LCC light grey to Leeds dark brown.

5: Reassembly

The cleaned, repainted and where necessary repaired framework of the first bogie was put back together whilst the other was being stripped down. The bogie was once again turned over and dry assembled, meaning that the components were trial fitted, and the tightening process was carried out only after all components had been proved to fit and operate correctly. Along with the wheelsets and motors, the suspension and braking system components, which had been refurbished in the Workshop, were dry fitted before the bogie was fully reassembled.

The whole process was repeated for the second bogie; it was completed more quickly, as many of the required parts had already been made while work was carried out on the first bogie. By the end of 2018, both the rebuilt bogies had been reunited with the body of the tramcar, and once again LCC No. 1 was mobile and could be moved around the Workshop to different locations.

Final machine work being done by Workshop volunteers on the new transom which had been made. (M Crabtree)

The first of the freshly-primed bogie frames, ready for parts to be refitted. (M Crabtree)

The first of the freshly restored components being dry fitted to the first bogie. (M Crabtree)

An indication of the complexity of the components being refitted to the transom on the first bogie.
(M Crabtree)

Many months of work and hours of meticulous refitting completed, the restored pair of bogies wait on the Museum's Depot fan to be reunited with the body of LCC No. 1. (I Ross)

Another team effort ensured the safe and successful reunion of the body and running gear. (M Crabtree)

Air supply system

1: Deconstruction

LCC No. 1 was one of only three London County Council tramcars to be fitted with air operated brakes, and the air system also provided power for features which were novel at the time: folding doors, retractable step, and a track sanding system.

All the pipework, along with its connection pieces and valves, was removed from the tramcar, with the team discovering that the pipes were a mixture of copper tubing and steel. There were five air reservoirs, two door motors located at upper deck floor level, and a compressor mounted in one of the driver's cabs.

Most of the network of pipes was installed beneath the tramcar and the lower deck flooring.

A challenging pipework puzzle, seen from beneath. It had to be understood and recreated to ensure successful operation. (National Tramway Museum)

The original air tanks on the south platform showed heavy signs of corrosion. (National Tramway Museum)

As with the electrical elements of the tramcar, there were no schematic drawings to help with understanding the pipework: where it all connected, what was original, and what had been modified whilst the tramcar was operating in Leeds. Working from photographs taken during the deconstruction process, the team laid out all the removed pipework and components on a wooden platform, thereby reconstructing the system so they had another 3-D physical reference for the reassembly.

When removed, one of the five air reservoir tanks was half full of a water and oil mixture, present at least since 1957.

Once all the pipework had been removed, it was laid out to be examined. (National Tramway Museum)

2: Analysis

As there were no original pipework drawings for the tramcar, the aim of this part of the project was to track where the air system went and what it used to do; to understand modifications made between the tramcar operating in London and operating in Leeds; to decide how those modifications could be reversed; and to assess how the reversals would affect the rest of the pipework. Once these questions had been answered, a new set of drawings was produced to record all the details, before any pipework was reinstalled.

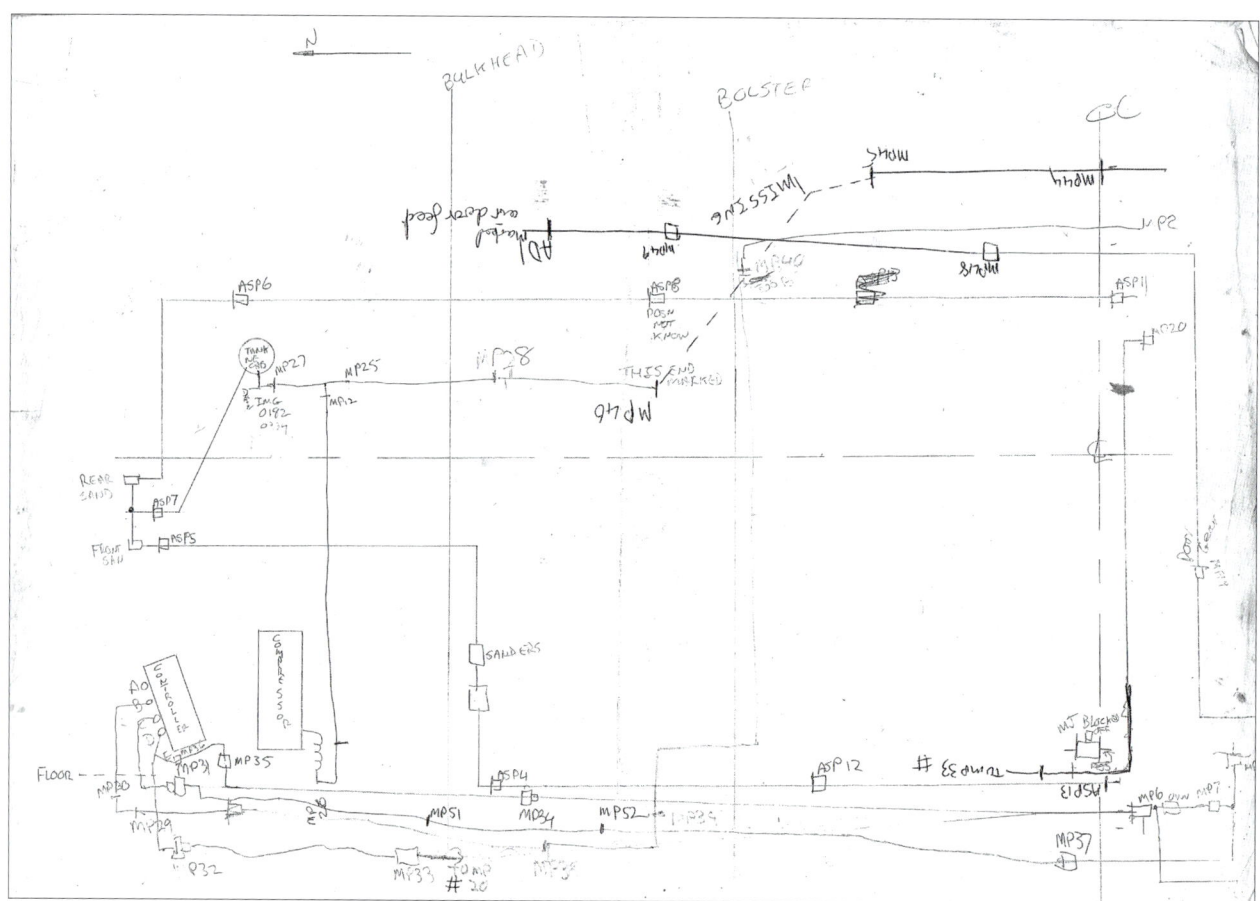

Workshop volunteer Andy Bailey figured out the puzzling pipework, producing drawings along the way to record progress and details for future maintenance and operation. (A Bailey)

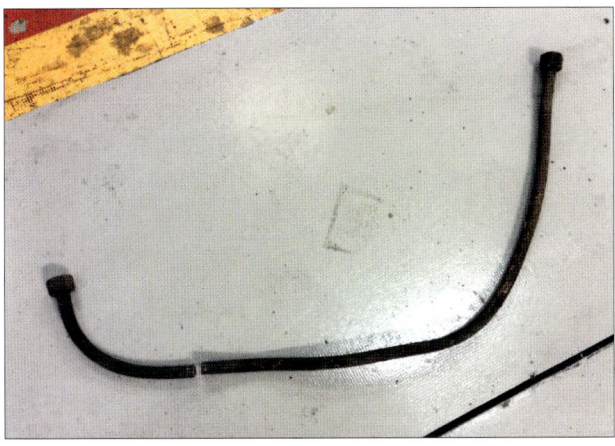
Many of the pipes were heavily corroded and unfit for further use. (National Tramway Museum)

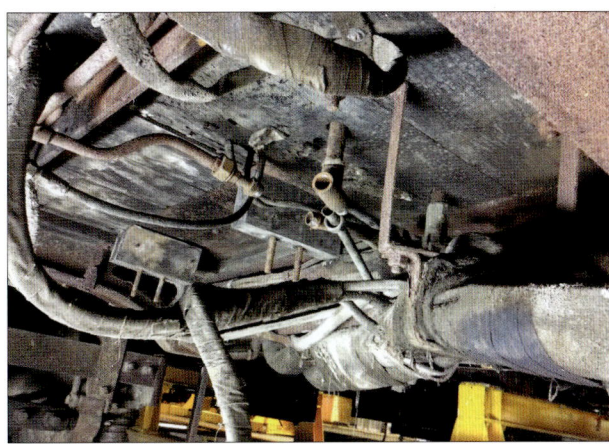
A section of pipework where a T piece was discovered, with no indication of what it was previously linked to or why it had been terminated. (National Tramway Museum)

The pipework consisted of a variety of long and short lengths. Some of it had been modified during No. 1's time in Leeds, when new air brake valves were installed to be compatible with the rest of the Leeds fleet; but most of it had been in place since the tram was built in 1932. Crushing or scraping of sections of pipework indicated that clearances around the bogies were inadequate in these areas. External corrosion was evident, and it was assumed that internal corrosion would also be present in the pipes.

There was evidence that some of the pipework had been rerouted at Leeds; there were some dead ends, such as T-pieces where an arm had been cut off. There were areas where no clear route or purpose for that part of the piping could be found, but in some cases there was evidence that it could have been part of the supply to the conduit changeover switching.

The compressor on the tramcar supplies four large reservoir tanks to maintain the pressure between 60 and 90 psi for the braking system; a fifth, smaller tank supplies a lower pressure system of up to 40 psi for the door motors. The compressor had been replaced at Leeds by a different type to that installed in 1932.

3: Treatment

As the condition of the original pipework could not be ascertained, the team could not say for certain that it would be safe to reuse in an operational system. External deterioration with rust could be seen, and the degree of internal corrosion was unknown. It was also possible that the steel had pinholing in thinned sections. Thus, the decision was taken to replace it all with new galvanised pipes. The number of connections would be reduced to minimise potential leakage.

When tested ultrasonically, four out of the five air reservoir tanks were found to be beyond repair due to corrosion from the inside. Five new tanks were manufactured by a specialist supplier and appropriately certified after testing. The team had to make a compromise in this area of the restoration as the sizes of cylinders available were limited, and none were a perfect replacement for the originals. Therefore, some of the newly commissioned cylinders had to be a slightly different size to that of the originals, in order to create a working system.

New air tanks were supplied by a specialist firm, and were installed on the south platform. (M Crabtree)

All valves were stripped down to basic components. New seals were fitted and tested after reassembly.

4: Leeds modifications

Leeds' engineers disconnected the controller mounted air brake valve, so that No. 1 could operate as a standard Leeds air brake tramcar. The compressor was exchanged for the standard Leeds Maley & Taunton type.

As with the cutouts along the body side, Leeds rerouted some of the tramcar's pipework to provide clearance over the bogies on the system's tight curves. With no need for conduit operation in Leeds, pipework and control valves for the conduit changeover air system were removed or terminated.

5: Reassembly

The work to restore a functioning air system on the tramcar involved using the original pipework which had been removed during the deconstruction phase of the project. As described above, this had been laid out as a giant three-dimensional model for use as a template for the new pipework.

The Maley & Taunton compressor was a modification, made when the tramcar was in operation in Leeds. (National Tramway Museum)

With new pipework throughout, the work on the air system needed to be co-ordinated with the installation of the electrical control gear.
The Conservation Workshop acquired a hydraulic pipe bender to facilitate this task. The first sections of pipe reinstalled were for the section of the air supply system linking the electrically driven compressor to the four reservoir cylinders.

In May 2017, tests with the renewed pipework took place, and parts of No. 1's system of pipes and valves had air going through them for the first time in over 60 years.

These initial tests with parts of the pipework and valves were undertaken using the Conservation Workshop's compressed air system. A sticky valve on the top of the north controller and a slight leak

A very different view of the underneath of No. 1 as the first of the pipework is restored. (M Crabtree)

on the rear brake cylinder were the only issues discovered during this first phase of testing, and these problems were resolved when the rest of the system had been fitted and the whole system was tested further. Brake cylinders for wheel and track brakes and the sanding gear were also fitted and tested.

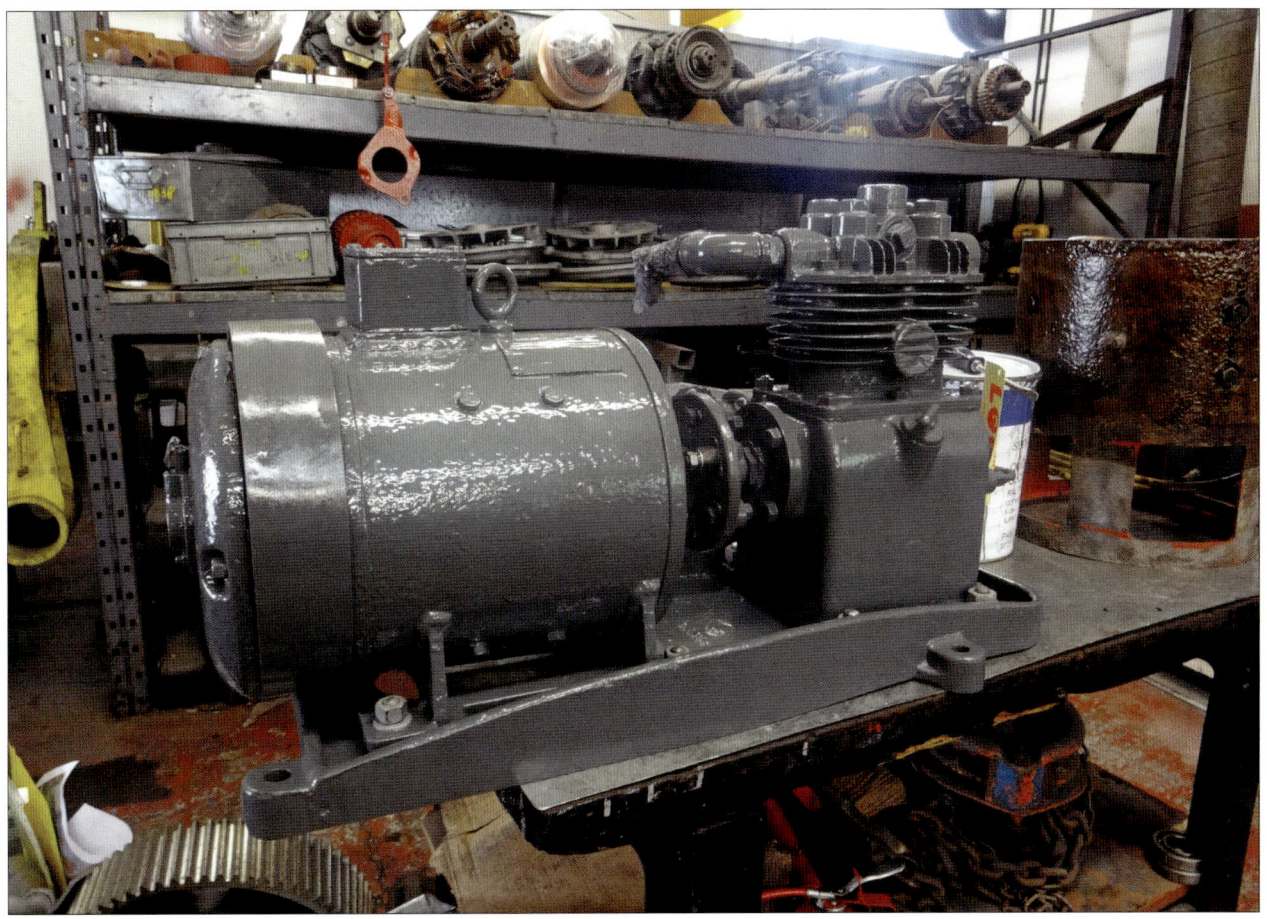

A Westinghouse compressor, of the type which LCC No. 1 would have had when originally built, was sourced from stores and dispatched to a specialist contractor for overhaul. (M Crabtree)

Initial trial fit of the folding step mechanism, to see if it would function with the air supply system which operates it. (National Tramway Museum)

The Quick Release Valve
LCC No. 1 has a quick release valve (QRV), which is positioned near the brake cylinder so that when the brakes are released, the whole system does not have to be exhausted. Instead, only the air in the system between the QRV and the brake cylinder is exhausted through the QRV, allowing the brakes to release much faster.

The Maley & Taunton compressor, installed at Leeds, was removed, and an example of the correct compressor, a Westinghouse type E.13, was located in the Museum's off-site store. This compressor was sent away to contractors for overhaul before being returned and installed on the tramcar.

As has already been mentioned, throughout the project the Workshop's mechanical engineers had been working to replicate the original folding step mechanism which had been featured on the tramcar. It was necessary for work carried out on the air system to connect with the mechanical operations of the step, as this was what powered the step mechanism.

Electrical supply system and associated mechanical equipment

1: Deconstruction

LCC No. 1 has the same basic traction and control units as the London Class HR/2 tramcars, with four Metropolitan Vickers motors and two Metropolitan Vickers OK 37B camshaft controllers. It also has the modification of an EMB air brake interlock head added to each controller.

Motors and Controllers
First generation tramcars had either two axles, each carrying an electric motor, or four axles, with motors on either two or four axles. LCC No. 1 is powered by motors on four axles.
The motors are powered by an electrical supply of 500-600V Direct Current. The amount of electricity delivered to each motor determines the speed of the tramcar and is controlled by changing the resistance in the circuit.
The mechanical device by which the motorman (driver) achieves this is called a controller. It is the large curved box with a handle on top which appears on both platforms of most tramcars.

The first of No. 1's controllers ready to be dismantled and meticulously recorded, so that refurbishment and asbestos removal could take place. (National Tramway Museum)

The traction motor carcases were stripped for repair and sent to a specialist sub-contractor for refurbishment.

The controllers were removed and separated into the air brake and electrical control components. Knowing that they would be a complex and challenging problem to deal with, the team decided that, as with the bogies, the second controller would be kept intact, for the team to use as a reference. In October 2016, therefore, the electrical team worked with a specialist contractor to take one of the controllers to pieces and remove asbestos from the components. Although asbestos has excellent heat resistant properties, it is a health hazard not appreciated in the 1930s, and is now a substance with legal restrictions on its use.

The removal was a painstaking process which took several days to complete. The internal wiring of the controller is formed from solid copper bar, each piece of which has a particular shape, to route between connections. Many of these had adhered to the asbestos backing of the controller. Each wire was individually removed, cleaned by the specialist to remove traces of asbestos, labelled and set aside for reinsulating. Once this process was complete, the main controller carcass and cover were removed from site for the bulk of the asbestos to be removed in a controlled environment.

The meticulous process of dismantling a tramcar controller, and the component parts that make up LCC No. 1's type of controller. (National Tramway Museum)

Resistance banks removed and ready for inspection. (National Tramway Museum)

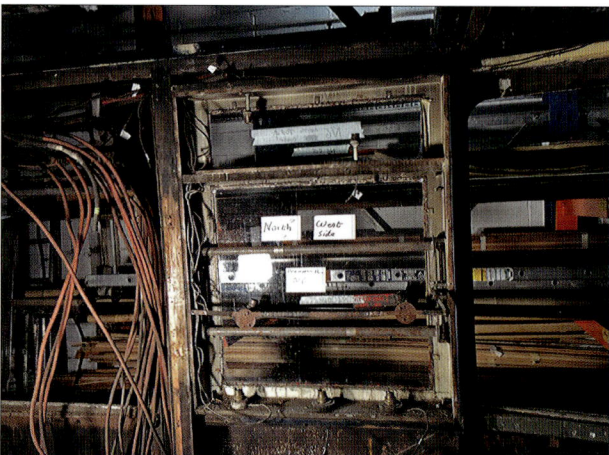

Like the pipework, the wiring was a complex puzzle to be solved by the Conservation Workshop's electrical engineers and volunteers, as no drawings of it existed. (National Tramway Museum)

The process would be repeated again for the second controller, at a later date.

The resistance banks, mounted under the platforms, were removed for refurbishing. Two circuit breakers were also removed, along with the main lighting fuse panel from below one staircase. Tracing the lighting circuits involved following the wires through the saloons and recording the position of the end connections, as no original wiring diagrams exist for the tramcar.

The traction cabling was all removed, with the major lengths extending along the tramcar's length, through the void in the lower deck saloon between ceiling panels and upper deck flooring.

2: Analysis

As with the pipework, there were no original wiring diagrams for the lighting systems of the tramcar. The team therefore had to follow a similar process: discovering where all the wiring went and what its function was, understanding the modifications made between London and Leeds, and considering the practicalities of reversing those modifications and the consequences for the rest of the wiring. Once this was clarified, a set of drawings was produced to record all the details, before anything was refitted onto the tramcar.

To comply with modern Health & Safety legislation, it was necessary to replace the old wiring with modern materials. Samples of all the different types of original wiring used were retained to assist in identifying the suppliers of the original materials.

For changing between conduit and overhead power supply, all LCC tramcars prior to LCC No. 1 had one or two manual changeover switches operated by the controller key. Probably because of its separate driving cabs, this tramcar had an air operated switch in one cab worked through two valves, with pipework running on the underframe [26].

The lighting circuits were related to those on earlier LCC tramcars, but had several additions which made the auxiliary electrics more complex. There were more lights for the saloons, the side service numbers, and the side destination displays, and also two unidentified lamp holders in cupboards under the stairs along with a two-way switch at one end.

A significant finding was that electrical staff at Charlton had twisted joints and used insulating tape apparently randomly and the wiring layout was not in a logical form. The probability is that as No. 1

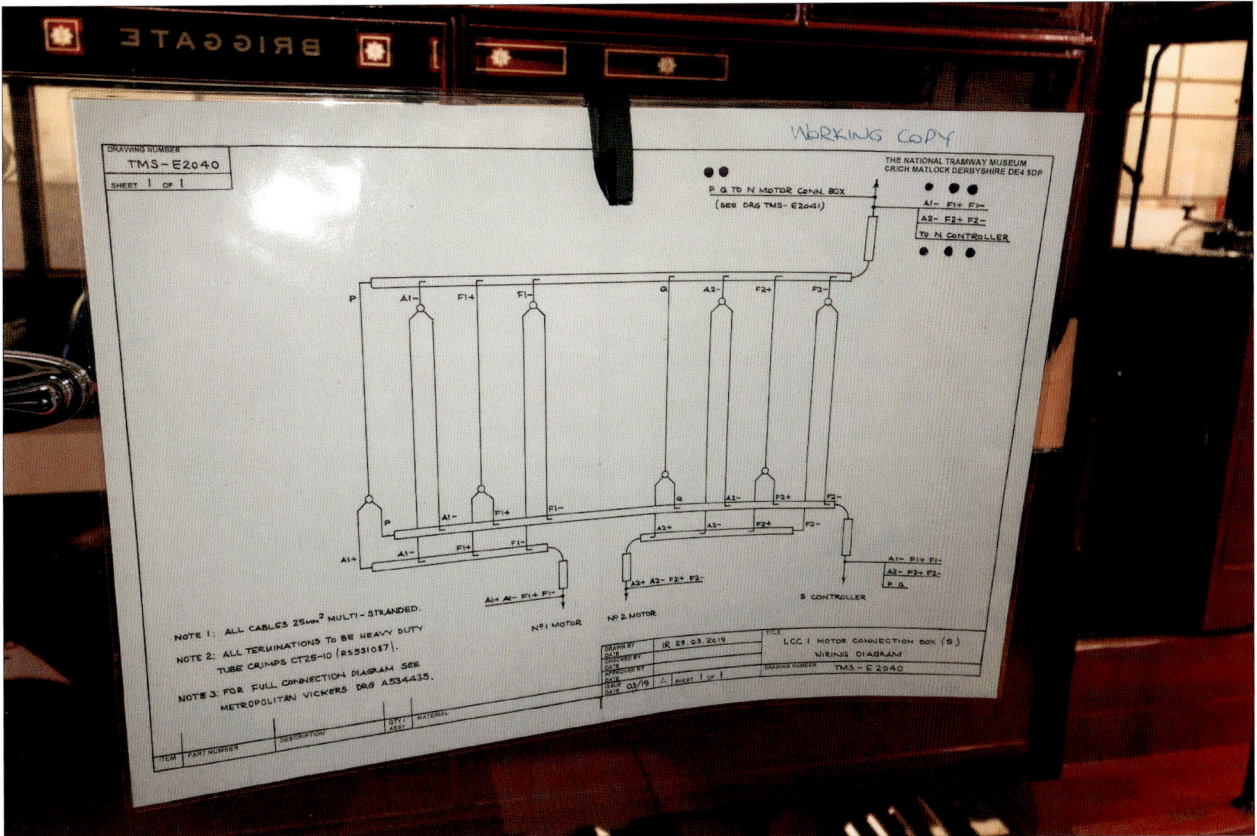

After years of work, the Conservation Workshop team have created new drawings which record the newly installed wiring. (National Tramway Museum)

The original sawdust-filled panel which was behind the compressor in the north driver's cab. It's thought to have been an attempt at soundproofing. (National Tramway Museum)

was a prototype vehicle, the wiring was put together and the circuits were worked out as construction progressed. It is understood that a wiring drawing may have been produced and placed in one of the cabs, only to be destroyed in a fire when in Leeds.

Evidence of a fire was found, after removal of one staircase, with burnt parts of both main power and lighting circuit cables. Some power cables had been cut and jointed. Much of the electrical insulation remaining in this area was damaged to the extent that it fell off when touched.

Some wires were found to have been threaded through the ceilings after the interior panelling was put in place.

Leeds' alterations or repairs included severing or removal of electrical circuits and components associated with conduit operation.

The controllers both had worn mechanisms and one showed signs of cables damaged by fire due to an earth fault. Some components, replacements for wear and tear of parts in service, were stamped LPTB, confirming ongoing maintenance in London Transport days.

Because the replacement compressor which had been fitted in Leeds was not to the original specification, the associated pipework required some rerouting. The compressor was partially enclosed in a cabinet with twin-walled sides, the gap between the walls being filled with sawdust to provide some acoustic insulation.

All the resistance frames were installed below the platforms. A section of one of them had been patched up to keep the tramcar serviceable. The position of the damage confirmed reports of a handbrake chain coming into contact with live parts of the frame.

The heaters, originally fitted behind skirting boards, had been taken out and disconnected. This had probably occurred in Leeds. The wiring and asbestos lining had been left in place.

3: Treatment

All the removed wiring for both traction and ancillaries was replaced with modern materials to meet current safety standards. It was noticeable that the newly installed wiring takes up significantly less space in the body side cavities.

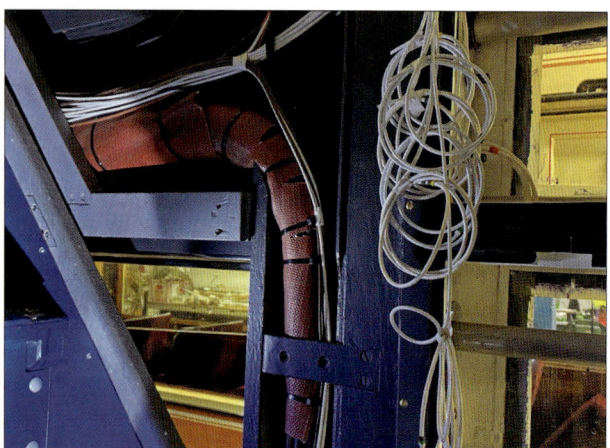

The lengthy process of installing new wiring throughout the tramcar can be seen here, as can the modern wiring being used in order to meet current Health & Safety standards. (D J Heeley/National Tramway Museum)

Fibreglass insulation replaced the asbestos-based insulation in the controllers. The electrical contacts were taken apart and renovated for reassembly. Worn mechanisms within the controllers were refurbished or replaced with new, purpose-made items.

Once the electrical and mechanical components had been reinstalled in the main controller housing, checks were made on the mechanical operation of the controller. Both controllers were renovated and reassembled in the same manner. One controller additionally needed repairs to segment carriers damaged by fire.

Resistance banks under the platforms were taken down and the asbestos-based heat resistant sheets above them were removed for disposal. The four resistance frames have been refurbished by EMB's successors. Three have been rebuilt using the original resistance strip and the fourth frame had a section that had to be patched up. The new strip was formed by the same method EMB have used since they developed the 'jointless resistor' in 1911.

The motor carcases were stripped; field coils were fitted prior to receiving the armature and then fully assembled and painted. The axles were then fitted [27]. The overhaul included new suspension bearing shells.

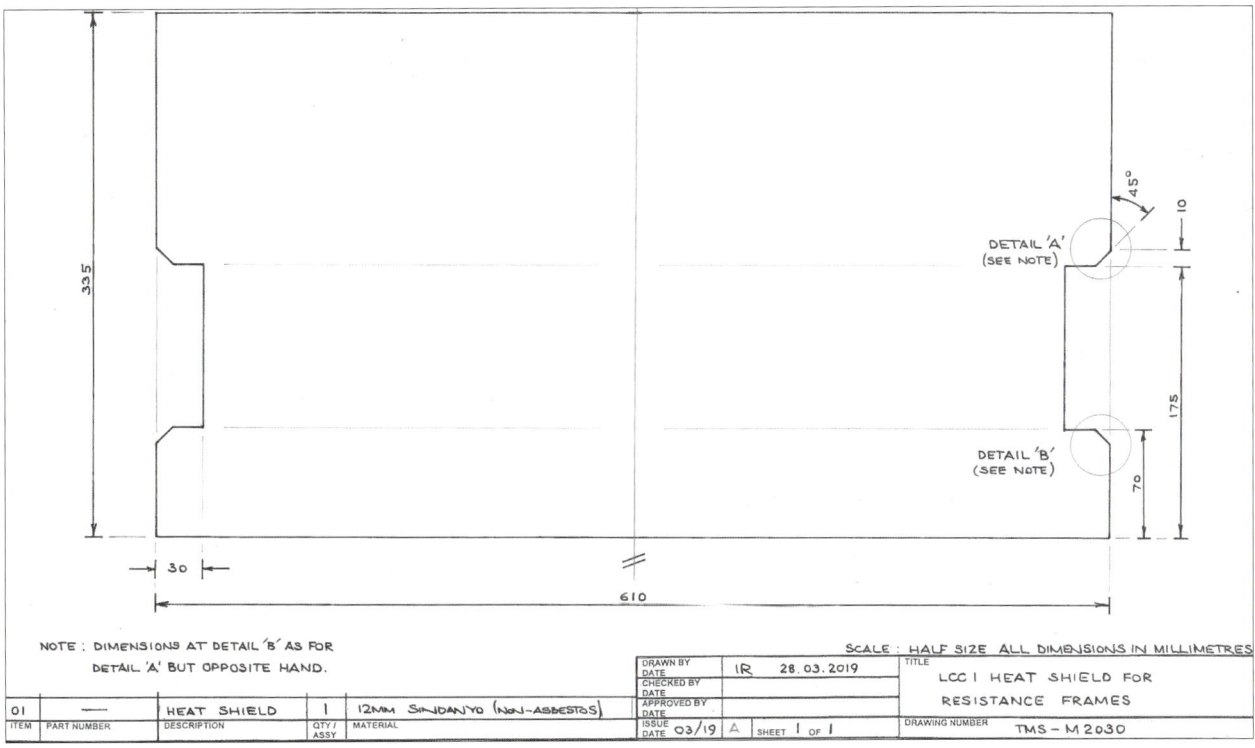

A number of new drawings have been produced during the course of the restoration project, including this one for a new replacement heat shield. (National Tramway Museum)

A superb illustration of LCC No. 1's two controllers - the first with components restored and refitted (left) and the second still to be restored (right). (P Whiteley)

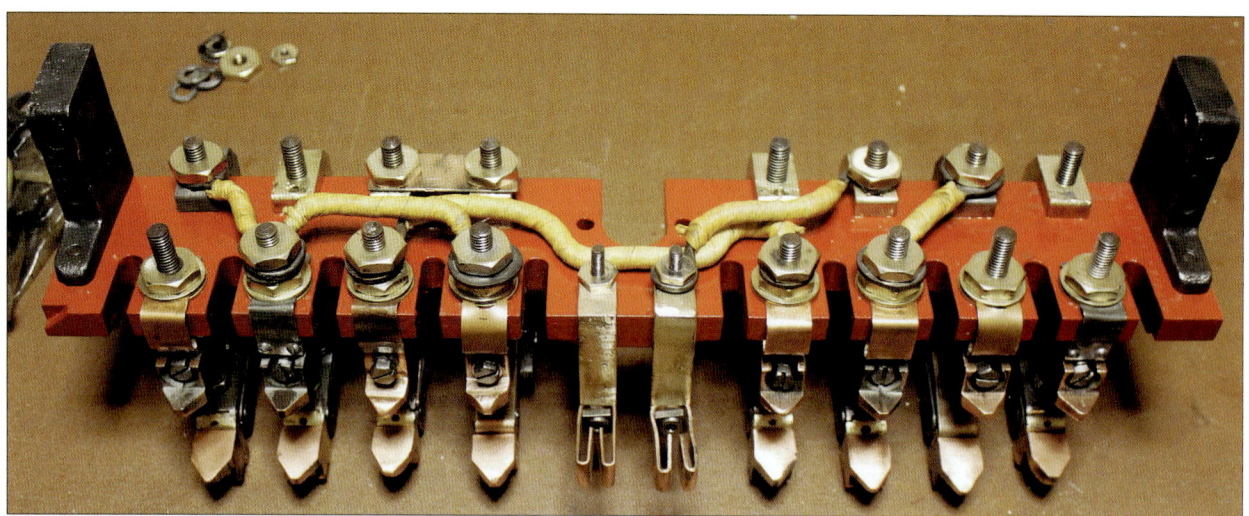

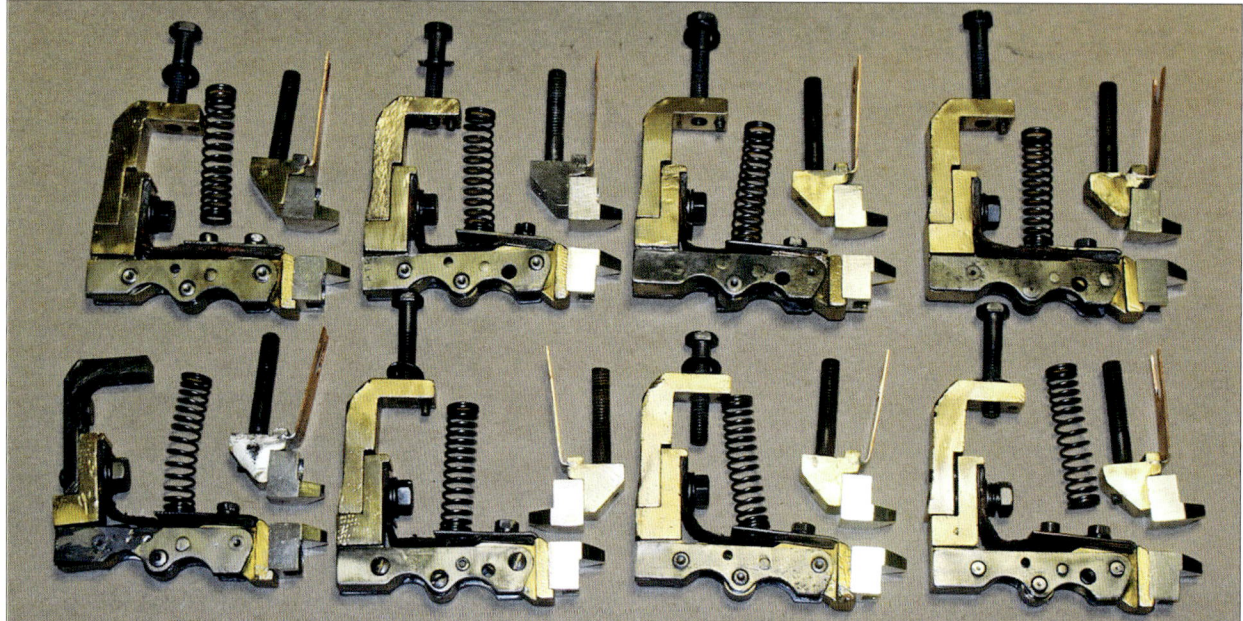

4: Leeds modifications

Looking back again at the Restoration Specification, several alterations made by Leeds have had to be addressed.

London's tramways' use of two power supplies, from overhead wires and conduit, caused complications. In LCC No. 1 this necessitated an extra pair of circuit breakers, one in each cab, and a supply changeover switch to disconnect the power to the trolley poles when running on conduit track and to reconnect them when required. Unnecessary circuits for conduit operation had been removed in Leeds.

LCC No. 1, when built, featured an additional fifth circuit breaker, located under one of the staircases. This also features on HR/2 No. 1858. It is an 'interlock' circuit breaker and is remotely tripped only if both controllers are placed in the 'forward' or 'reverse' positions at the same time. It is thought to have been used to protect the vehicle when fast changeovers, involving a second driver taking the tramcar, were being made at termini.

Close-up views show the huge amount of restoration work that has been completed on the component parts of the controllers. (P Whiteley)

The additional circuit breaker was removed in Leeds. A replacement was taken from stock; it came from a Glasgow roof-mounted circuit breaker. Evidence of the original wiring for this circuit was found.

The tramcar also had an air-operated changeover switch in the north cab, worked through two valves and pipework running on the underframe. Leeds removed most of the changeover switch air system leaving one valve in one cab and two flattened pieces of copper pipe in the other.

Leeds made several alterations to equipment over and above the conduit side of the traction circuits. The negative fuses in all the lighting circuits were also removed, but their positions in the fuse panel remained. Saloon heaters were also removed, leaving cut wire ends. Additional wiring was found running between the two controller blowout coils, placing them in parallel. The reasons for this are unclear, but it does appear to overcome the removal, perhaps in error, of some of the circuits mentioned above.

At the Museum the traction and lighting cabling was replaced and reconnected to include conduit operation configurations.

When Leeds fitted their standard Maley & Taunton air compressor to replace what was a Westinghouse unit, some wiring for the original compressor was chopped off and left in place in the cable ducts.

The traction voltage bell system had been split into two parts, leaving a small number of bell pushes working the London bells with a battery voltage buzzer system added.

New trolley bases, trolley poles and trolley wheels have had to either be made from available material or machined from new, as these items were all replaced in Leeds with a Fischer bow collector.

5: Reassembly

New cabling to the required insulation standards has been installed on all circuits to meet current electrical regulations. The main traction wiring connects the two controllers, the four motors, the four resistance banks and the four track brakes. The tramcar now has thick black wiring throughout.

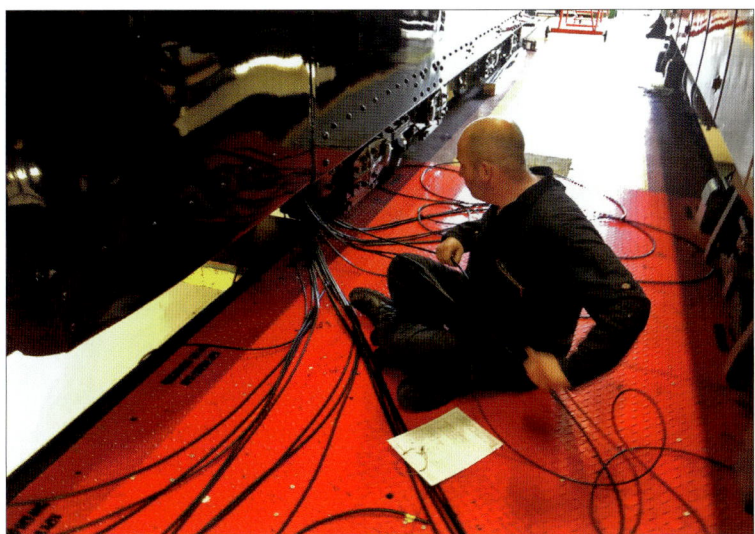

Conservation Workshop electrical engineer Jason Watkins working through the installation of the traction wiring. (D J Heeley)

A close-up view of a resistance bank installed under the tramcar, and the wiring feeding into it. (National Tramway Museum)

Most of these cables are run within the lower deck body sides and within the ceiling void; they had to be threaded through carefully to avoid cable damage. As we've seen earlier, this installation and testing of the wiring running through the voids meant the Workshop's coach builders had to wait to install the ceiling panels which would cover these areas and conceal the wiring. Once the cables have passed through the floor to the underside of the tramcar they run into a protective hose and, grouped into bundles, pass to the controllers or resistance banks.

The controllers on LCC No. 1 are very complex, and therefore more cabling is present than on some earlier tramcars. This is because the motors are controlled individually in each controller rather than being connected in motor pairs and switched as pairs. The reassembled controllers had to be adjusted to give correct sequencing and smooth operation.

Once installed, the traction cables needed to be routed into a motor disconnection box, before cutting back, labelling and terminating. These terminations were originally done with soldered 'T' joints in the cabling, but this is not a current good practice; mechanical, screwed, connections have been used instead. These boxes will be hidden from view, located underneath each of the staircases.

The black traction wiring can be seen here bundled together, whilst the brown and white wiring for the electric bell and lighting circuits continues to be installed.
(National Tramway Museum)

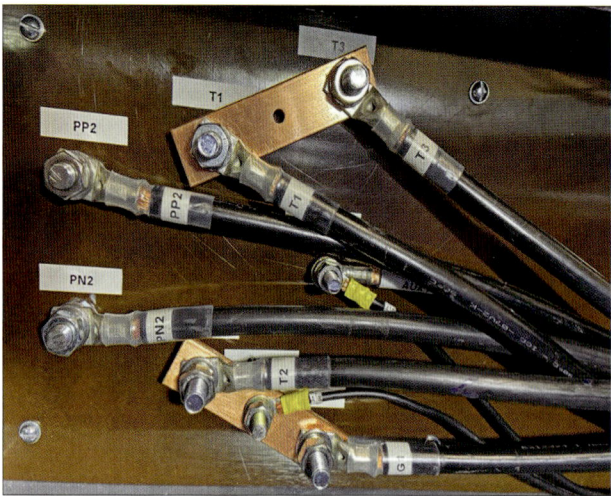
A close-up view of the non-functioning wiring for the air changeover switch demonstrates the great care the electrical team took to identify all wiring for future maintenance and operation.
(D J Heeley)

Not all the wiring removed from the lighting circuits was found to have a purpose, and for the restoration a Curatorial and Workshop compromise was reached: as long as all lights operated in the correct manner, this would be acceptable. White and brown wiring was installed for the lighting and electric bell systems to distinguish the two sets of circuits.

Wiring and connections for an air-operated changeover switch, which would have allowed power to be swapped between overhead supply and conduit, were installed as part of the reassembly work. The connections were then terminated so that they don't function; however, they are present if in the future such swapping between overhead and conduit operation were to be feasible.

Conduit circuits and plough carrier components have all been reinstalled. This work included sourcing three additional circuit breakers from the Museum store which were overhauled before installation.

Workshop volunteer and LCCTT Chairman Ian Ross patiently awaits the go-ahead to proceed with the first powered test of LCC No. 1. (P Whiteley)

Testing of circuits, operation of equipment and checking wiring connections to the motors for correct rotation was undertaken before applying traction voltage to the tramcar.

On 22 April 2022, all eyes at the Museum were on LCC No. 1 as the Workshop team prepared to discover whether or not the tramcar would move under its own power for the first time since 1957, when it ran for the last time in Leeds.

In the safe hands of Workshop volunteer and LCCTT Chairman Ian Ross, LCC No. 1 cautiously left the Workshop under its own power. The hours of painstaking work which had resulted in the reinstallation of the electrical system had allowed the tramcar to operate once again.

Testing and commissioning

Final testing, and then commissioning, are staged processes undertaken over an extended period during the final stages of restoration. As explained earlier, it is necessary to undertake some testing with the tramcar partially completed as components can become inaccessible later – for instance, the main traction wiring is sealed in situ within the upper deck ceiling voids, so must be tested before the ceiling panels are installed permanently.

Trolley pole connected to the overhead line, LCC No. 1 can be seen exiting the Workshop under its own power. (P Whiteley)

Under the watchful eyes of various Workshop staff and volunteers, LCC No. 1's first powered foray out onto the Depot fan was a success. (J Dignan)

Many components are first tested as sub-assemblies, with items such as controllers and bogies being tested away from the tramcar before being integrated into the vehicle.

Once the body of LCC No. 1 had been placed back onto its bogies, it was towed around the tightest curves on the Museum's track, which are on the Depot fan, to undertake an initial check of the restored body's bogie clearances. This revealed a number of areas that required alteration to avoid clashes. Such is the complexity of LCC No. 1 that, even with adjustments made by the team, it will still be restricted from operating on certain parts of the tramway because the team have been unable to obtain the required clearance between components.

Once the team were satisfied with this stage and the air piping was completed, the body-mounted brake gear was connected to the trucks. The tram was connected to the Workshop's internal compressed air supply. This allowed extensive testing of the air systems, to fix any minor leaks in the pipework and to test and adjust the brake operating mechanisms and the air operated folding step system.

A pipe fouls whilst testing is conducted on clearances around the Museum's Depot fan. (D J Heeley)

LCC No. 1 on one of the tightest curves on the Museum's Depot fan. (D J Heeley)

Whilst all of this was ongoing, the electrical work was proceeding towards completion. As with all the disciplines, hundreds of hours had gone into the identification of the original wiring arrangement, so that it could be redrawn in preparation for installation. Once completed, the fresh set of drawings was used and everything marked off to ensure wires were correctly labelled and the correct number connected to each terminal. Circuits were then checked for continuity and insulation from the metal body.

Only after all this testing had been undertaken was it possible to energise No. 1 and test its circuits. This was done a stage at a time. First, the automatic trolley changeover circuit was tested. During this phase, a temporary light batten was affixed to one end of the tramcar to indicate power was available. Next the traction circuits were tested, resulting on 22 April 2022 in No. 1 moving under power for the first time in over 60 years.

A common difficulty with tramcars is getting all the motors to turn in the correct direction, and despite the Workshop's best endeavours it was found that one of the motors was trying to turn in the opposite direction to the other three. A quick return to the Workshop, where the cables to the offending motor were swapped over, easily corrected the problem.

Over the following months, the electrical testing moved on to the lighting circuits, and at last, with several circuits operational, the very satisfying view of an illuminated LCC No. 1 was again to be seen in the Workshop.

The lighting circuits are arranged in two distinct groups, the interior lighting being one and the exterior lighting and destination boxes being the other. The latter group was the last to be connected, ending the electrical testing phase.

An illuminated LCC No. 1 seen in the dimmed light of the Conservation Workshop. (P Whiteley)

Once the system had been proven, the remaining coach work, as described previously, could be fitted. The last electrical connections to be made were the electro-magnets for the track brakes, which had been shorted out until the last stages of restoration.

Whilst the testing of the various systems and final electrical connections was progressing well, one major problem was causing the Workshop team some consternation, ahead of releasing No. 1 for further testing – the compressor.

Fuses were blowing when the compressor was started against a load. It seemed that LCC No. 1 had one final challenge for our engineers, and a tricky one at that.

Trying to identify what was causing the issue would not be an easy process, as the team needed to eliminate a range of options: was it the wiring that had been installed on the tramcar, or were the fuses faulty, or was it a specific issue with the compressor itself? Whilst the compressor had had a full overhaul early on in the project, the nature of historic equipment is such that it can still catch you out with another fault further down the line.

After investigation by external contractors and further work by the Museum's engineers, the fault was identified as being with the brush gear in the compressor. The process of identification and rectification would delay testing by an additional three months.

Once the compressor was reinstalled at the beginning of July 2023, the trolley poles were tensioned, and the compressor carefully tested.

In a significant milestone moment, and as a reward for overcoming what they hoped would be No. 1's final challenge, the Workshop team ran the tramcar for two round trips on the Museum's main tramway from the Depot crossover point to Town End terminus on 8 July 2023. Rolling Stock Engineer Peter Bird described this as a "…satisfactory initial test with no significant defects evident…".

LCC No. 1 during its round trip to the Museum's Town End terminus, 8 July 2023. (D J Heeley)

At this stage No. 1 was being exclusively driven and tested by Workshop staff and volunteers. These would then continue the running-in process, mostly on days when the Museum was closed or after opening hours, ensuring a quieter, controlled environment in which to observe how the tramcar was functioning.

Only when the Workshop team were confident that No. 1 was functioning correctly could the testing and driver familiarisation responsibility be shared with a set of qualified driving instructors from the Museum's Traffic team. These instructors would later instruct other drivers from the Traffic team on how to drive No. 1, and on any practices that are unique to the tramcar.

During this period, any shortcomings would be highlighted in the risk assessments and jointly reviewed and signed off by the Workshop and Traffic teams.

Accumulating a decent amount of mileage is really the only way to test a tramcar in preparation for continuous operation. To be able to do this, the Workshop team needed to complete a commissioning examination, in line with the Museum's usual tramcar maintenance regime. Every operating tramcar has this 'MOT' before the start of every operating year. Once commissioned, No. 1 could operate when the Museum was open.

Repeated inspections follow the commissioning, enabling minor adjustments to be made in response to any problems identified. In addition, the Restoration Specification is reviewed, final snagging is undertaken, and gradually the pool of drivers is increased, to help to accumulate the required amount of mileage.

Only after the completion of all these steps could LCC No. 1 be declared ready to return to regular operation.

…and finally:

After nearly eleven years of dedication and hard work from a team of 46 staff and volunteers, No. 1 is once again ready to ride the rails and carry passengers in luxurious comfort.

What started as an ambitious four-year project to physically restore the tramcar has evolved into something much more challenging and time-consuming, with No. 1 spending just over nine years in the Conservation Workshop.

Although contingencies were built into the original programme, and despite the 18 months of preplanning, scoping and examining work, there were inevitably factors beyond the team's control, and surprises from the tramcar itself which would change the course of the restoration.

The team was challenged by a lack of contractors for specialist machining work, the unavailability of critical materials and components, and the lack of commercial facilities for specialist processes, for example chrome plating. The difficulty involved in sourcing contractors or finding solutions contributed to the extension of the restoration programme.

These problems also remind us that the nature of the conservation/restoration sector is entering a phase of change. More and more traditional engineering trades and processes are fast disappearing, with some having already vanished. The work of the Museum's Conservation Workshop in restoring and maintaining the Nationally Designated tramcar collection cannot be underestimated in its importance and the part it plays in sustaining as many of these traditional skills and processes as possible.

The project has seen people come and go, depending on the time they could commit to the Museum, as well as the natural progression of people moving on to other roles beyond the Museum. Nearly 25,000 hours of staff and volunteer working time has been contributed to the project. We've also seen the passing of colleagues during the decade-long project, but their contributions live on as a legacy through the achievement of restoring LCC No. 1 to an operational condition.

Although more than 90 years separates them, the threads of time have connected the original builders of LCC No. 1 with today's team. The team have throughout the project maintained a deep respect for the history of the tramcar and great diligence in recording all they found, with the aim of replicating and restoring No. 1 to, as nearly as possible, its original state.

Whilst the Workshop team felt their connections with the original builders, the Curatorial team, through the many artefacts found on the tramcar, were able to discover and share moments of time lived by those who rode and worked on No. 1. We had expected to find some evidence of these moments, but the possibility of finding such a range of artefacts was never in our minds when we first set forth. To be able to connect the different chapters of LCC No. 1's journey from initial build, to operation in London, through its move to Leeds and finally to its life as part of the Museum's National Collection, is a rare occurrence.

As work on the restoration extended into 2020, progress was interrupted by the unexpected effects of the Coronavirus pandemic. Necessary restrictions as a result of lockdowns and, later, the required separation into 'bubbles' to ensure potential availability of Workshop staff resulted in virtually no work being undertaken on LCC No. 1 for over a year.

This can be viewed with a degree of irony as the original construction of LCC No. 1 was also halted, with no work undertaken for a period. However this was only between September and December 1931 and was due to the poor 'health' of the London County Council which had to make some rapid budget cuts, not caused as in this case by a public health pandemic.

I once asked Workshop volunteer Ian Ross how he approached the massive challenge of working out the wiring on LCC No. 1. His reply was "patience and ingenuity", which feels like a common theme for all disciplines on the project.

Inevitably there have been frustrations within the team, when deadline pressures have occurred or when interconnecting work packages have been delayed, perhaps by one of the disciplines struggling with supply/contractor issues. A project as long in duration as the restoration of LCC No. 1 is bound to go through its high and low periods; but through it all, the commitment, dedication, patience and perseverance of the Workshop team has enabled them to overcome all the challenges, and to produce work of the highest standard, finally fulfilling the required Restoration Specification.

As this chapter in No. 1's story is concluded, a new one begins with the tramcar's launch back into operational service as part of the Museum's demonstration fleet, to once again play its part in the lives of all those who ride on it.

Finally, everyone who has been committed to delivering this restoration during the past decade is now part of the continuing history and story of LCC No. 1 - Bluebird.

LCC No. 1 in the Workshop in early March 2023 after the decorative gold lining and fleet number had been added. (National Tramway Museum)

Appendix 1

The Tramway and Railway World and *The Electric Railway, Bus and Tram Journal* coverage of LCC No. 1, May 1932.

The Tramway and Railway World and *The Electric Railway, Bus and Tram Journal* in May 1932 both described the new tramcar with virtually word for word reports based on the information supplied by the Rolling Stock Engineer, Mr G F Sinclair:

"The new car is based on the experience gained from the HR/2 type. It is constructed to larger overall dimensions in regard to length and width, the design having been planned to take advantage of all permissible clearances as between vehicles of the same type passing on adjacent curves.

Exterior Finish A novel scheme of decoration planned on camouflage lines to accentuate the stream-lining has been adopted for painting the exterior of the car. The colours are Royal Blue with a groundwork of Ivory White, which is enriched at certain points by the addition of gold lining, whilst in accordance with practice protective parts continuously liable to traffic damage are finished in black so that they may quickly be restored. Polished stainless steel or chromium plated metal for the window rails and panels, also for the door commode handles and traffic auxiliary fittings, add attractiveness to the finished car. The under gear and trucks are so treated that apart from the general outline they are rendered much less obvious than is usually the case.

Destination Indicators
A bold screen is brilliantly illuminated at night to show both the service number and the key route points of the journey, and occupies at a convenient width the whole of the space between the main window of the driver's cab at each end and the corresponding window in the upper saloon. In substitution for the reversible traffic service and streamer notice boards, generally carried in brackets on the car-sides, a screen is provided for each side of the car adjacent to the staircase, which will be readily seen at the leading end on the near side by passengers waiting to board.

Seating
The seating provides for 66 passengers as against 74 in the HR/2 class, 28 being accommodated in the lower saloon and 38 upstairs. All the seats are staggered, thus facilitating movement in the gangways. Transverse seats are reversible. The intention in reducing the accommodation to 66 is to provide more comfort for the midday passenger. Compensation is also to be found in the fact that there are 56 sq. ft. of standing area for use at peak traffic periods, while the plan of construction enables the staircases to be placed inside of the car body, this giving facilities for ascending to the upper saloon at each end of the car. "There is a reduction in longitudinal seats. In the lower saloon these are a seat for two persons facing the side of each staircase, while upstairs seven persons can rest on the seats occupying the half-curve opposite the railing guarding the top of the staircase. By adopting the 'armchair' type in each saloon each passenger is given the comfort of an individual seat. The upholstery is in moquette, with edgings of hide or grained Rexine, a slightly different design being used for distinction between the upper and lower saloons.

Internal Finish
The internal decoration of the lower saloon is carried out on very attractive lines. A proportion of the woodwork has been left in its natural grain, filled up and varnished to a fine surface. The moquette upholstery of the seats has an antique blue foundation relieved by occasional overlays of salmon pink between two shades of fawn, whilst the embossed hides used for the fronts and ends of the seat cushions and the piped edges of the back rests are in a much deeper shade of blue with a little black in the grainings. To form the ceiling of the lower saloon a semi-clerestory type of construction has been adopted. This provides conveniently for the accommodation of the new type of semi-concealed lighting used, and has enabled the most convenient angles for distribution of the light to be adjusted as necessary. "The curved cove panels at each side of this ceiling, together with the arch panels and overhead bays, are finished in dead matt white stippled enamel, relieved, in the case of the coves, by panelling out in bright polished stainless steel mouldings, made to match the lamp frames, and returned at each edge and for the full length of each shallow clerestory rail. All remaining interior woodwork below the line of the springing cornice for the ceiling has been finished with cellulose lacquer, in an attractive shade of lustrous Chinese blue. At suitable points, this is further enlivened by a judicious use of polished stainless steel

edging and dividing strips, whilst in such positions as the space available will permit, panels and inserts of specially chosen hides or suitably grained Rexine of attractive shades to tone with the general ensemble have been utilised.

"A convenient space on the inner side of each staircase casing has been recessed and fitted with a plate glass panel framed in stainless steel for exhibition of the Council's posters.

"The floor covering is in linoleum of a deep rich shade chosen specially to throw in higher relief the colours embodied in the seat coverings and general finish, which effect has been pleasantly amplified by the attractive bronzed finish of the protection grids for the several heater units fitted against the skirting rails, for the full length at each side of the saloon. With the exception of these grids, all remaining metal fittings in this saloon, together with the staircase hand-railings, are made in polished stainless steel or in chromium finished.

"The structural work and panelling of the staircases and entrances is neatly finished in a practical manner for frequent easy renewal against extreme wear and tear, to tone generally with the planned effect of the main saloon.

"For the upper saloon, the work of decoration has been much simplified as a result of its spaciousness from end to end, practically without structural obstruction of any character other than the narrow pillars used between the windows for supporting the roof. The ceiling for this has accordingly been planned as a continuous curved panel, in one piece, from end to end, and of even surface, with the exception of the brightly polished hit-and-miss grids for the air extracting ventilators. This panel is supported at all edges by a wide cantilever cornice of simple formation, the underside of which serves also for reception of the semi-concealed lamp fittings. Stainless steel mouldings are used, a contrast to the flat white stippled surface of the ceiling.

"The upper saloon being used principally by smokers, a stout type of non-scratchable Rexine cloth in a shade closely approximating to the general colour tone of the lower saloon has been adopted, whilst treatment of the bannister rails and guard screens at the stair-heads matches this. The general style and finish of the seating is similar, with minor modifications, to that of the lower saloon, although with a view to variety the colours in the moquette coverings are not identical.

"The floor covering is of the same character and colour as in the lower saloon, whilst the grid covers for the separate heater units are similarly finished.

Staircases

At each end access to the upper saloon is provided by an easy staircase of ample width, the two bottom treads being so placed that the first one is parallel to and immediately opposite the main car entrance. Thus the incoming passengers divide into separate streams for the two saloons. A casing encloses the staircase and serves also to accommodate sand storage hoppers and switches and fuses needed for certain lighting and other circuits.

Heaters

Tubular heaters are provided just above the floor level along each side of both upper and lower saloons. The heaters are operated from the line supply and so arranged that sections may be varied to produce the desired temperature.

Windows and Ventilation

Each side of the lower saloon has four large windows, the two end ones arranged so that they can be dropped to practically the full depth of the framing while the two centre ones are of the fixed type, each with a small permanent opening for ventilation purposes which is overlapped at the top by twin glass louvres in metal frames, all so arranged that adequate ventilation can be obtained during bad weather without opening the windows beyond the extent of the louvres. To provide for end to end ventilation when necessary through the driver's cabs, the bulkheads of these are provided with laminated push-and-pull metal inlets. In the upper saloon, the whole space, except the framing uprights, between the bottom elbow rail and the underside of the ceiling is glazed, so that passengers have full opportunity to witness passing items of interest en route. The four main windows on each side are of similar type to those of the centre spaces of the lower saloon, with the exception that weather-protection is provided by the roof in place of glass louvres. As the upper saloon is primarily intended for smoking passengers, four Colt type ventilators with large adjustable hit-and-miss outlet grids have been fitted to the centre upper ceiling panel. Wherever practicable the glass used for windows and similar purposes is of non splintering type and British manufacture.

Lighting

Artificial lighting in both the main and upper saloons is provided on a generous scale, each unit in a separate reflector concealed by a panel of decorative glass. The reflectors have been designed for each separate case to provide an even and comfortable flood lit effect over the full area of each saloon between the seat and the ceiling cornice levels, and have been made component parts of the ceilings, without resultant shadows or interference from projections.

Constructional Features
Experience in metal construction gained during the production and service of HR/2 vehicles has been utilised in designing the body, which consists of a rectangular steel box of which the sides form deep self-contained truss girders, and comprise the equivalent of all normal side panelling complete with the sole-bars. The component structures are continued vertically above the normal level of the lower saloon roof to the height of the bottom rails supporting the upper saloon window frames, at which point the all-steel structure is completed. With a few exceptions, necessary to provide in emergency for quick removal and replacement of specific detail parts, the steel-plates and sections comprising this main structure are secured together by electric arc welding. To ensure the maximum practicable reduction in weight of the upper part of the vehicle, the structural frames for the windows in the top saloon are cast each as a complete unit from modified Alpax aluminium alloy, and bolted into position relatively to the top edges of the main steel side structures, the under side rails of the actual roof and to each other, in such a manner that the whole can separately be assembled for mounting on the main structure as a separate self-contained unit.

"As a result of adopting this scheme of car body construction a departure has been made from the general curved formation of the main saloon roof sticks or carlines, which in this case consist of mild steel angle and channel bars laid straight from end to end and transversely, so that imposition of the upper deck passenger load will have the effect of restraining rather than bulging the car sides, which is the general tendency on cars built with upper deck floors of conventional curved pattern.

"To the top faces of these flat carlines is secured a sheet of structural plywood, approximately 5/16 in. thick in one piece, from end to end and side to side of the car body framing, to which it is also secured at all edges, the necessary openings for staircases, etc., having been cut after its assembly. By this means an exceptionally solid and light flooring is formed to the upper saloon in such manner as also to provide a particularly effective horizontal truss or bulkhead to all vertical and longitudinal side structural members of the body, whilst the spaces previously occupied by solid packings needed for levelling up purposes to provide a level floor in the upper saloons are made available for accommodation of lighting cables, ventilation ducts and miscellaneous apparatus.

Roof
The roof top consists of a sheet of structural plywood, in one piece from end to end and side to side, where it has been shaped to the requisite curvature by steaming under hydraulic pressure, stiffened by laminated timber and other carlines, and finished at each end by aluminium castings which furnish lateral stiffness and form the overhanging cowls needed for weather protection of the end windows and indicators.

Lower Deck Floor
In order that the lower saloon floor should continue at an even level from end to end of the finished vehicle, including both the motorman's cabs and passengers' entrance lobbies, at a height above rail level not exceeding that of the under-hung cantilevered platforms on existing cars, whereby the vertical space available is considerably restricted, special attention had to be given to the selection and design of the underframe transverse members, especially the main supporting body bolsters and the intermediate headstock supports for longitudinal platform and cab bearers. As a result the body bolsters have been compounded from standard sections of rolled steel joists and channels, embraced by top and bottom flange plates welded and riveted together into the form of a wide and shallow box girder. The intermediate headstocks have been prepared as steel castings, shaped to form gussets at the junctions with the body sides and sole-bars, and with flanged recesses to support the platforms and cab bearers, which, within certain limits, are free to slide therein, thereby dispersing shocks resulting from minor traffic conditions or the like. Wide gusset plates are fitted as stiffeners to the junctions of all principal longitudinal and transverse members of the main floor or underframe, which also serve as foundations for the lighter angle and tee bars used for supporting the removable hatches giving access to the motors and current collecting (conduit) plough, which latter is supported, with freedom to traverse from side to side, in rolled steel channels.

Flooring
The flooring consists of tongued and grooved boards, mostly of an Empire equivalent to pitch pine

timber, and in some locations of oak, which are bolted and screwed direct to the steel framing with stout intermediate timber packings where necessary. Above this is provided for sound-absorption purposes a layer of Insulwood and the top surface finished with stout hard battleship linoleum, of a shade complementary to the interior decoration of the saloon which at certain points, particularly in the motorman s cab is further protected against damage by layers of asbestos and steel plates. Similar materials and methods (with the exception of supplementary layers of asbestos and steel plates) are used for finishing over the structural plywood foundation of the upper saloon floor.

Entrance and Exit Doors

The entrance doors, combined with safety lifting steps and side life-guards, are air-operated by the driver. Under normal conditions the car will operate with the rear doors open and the corresponding step let down, whilst the doors at the leading end will be respectively closed and the step 'up' to ensure statutory clearance space above the life-guard tray at the leading end. The normal entrance space between the side doors and staircase casing at the front end will thus be available for the accommodation of standing passengers. The doors are in pairs, each right and left unit being articulated for overfolding and economy of space when opened. The opening and closing is effected by a double-ended air engine mounted on the upper saloon floor above the doorway and coupled to the hinge post of each right and left hand unit. The bottom end of the inner hinge post is furnished with gearing to engage the horizontal shaft of the lifting step and ensure that the operation of this either upwards or downwards synchronizes with the closing or opening of the doors.

Bells and Gongs

Communication between passengers and the driver or conductor is furnished by electric bell pushes, operated direct from the main line supply, located in convenient positions adjacent to the ceilings, staircases and doorways to serve the requirements of rapid services with frequent stops. To provide for urgent or traffic signals between the driver and conductor a supplementary service of atmospheric air bells is included, with signalling points to and from each cab and at each entrance. The traffic warning gongs, fitted under the platforms, are arranged for electrical operation from the main line supply by means of a pedal near the control equipment in each cab.

Safety Devices

Fitted to each dash plate is a special type of headlight, with which a focussing device has been provided to permit of the unit being used as a fog-lamp when needed. A near side overtaking traffic warning lamp signal is coupled to and operated automatically from the brake apparatus.
"Another safety device consists of a double screen wiper, air controlled, at each side of the main cab windows.
"The statutory police and overtaking traffic lights at the extreme width of the structure are approximately at the floor level of the upper deck. The change from red to white and vice versa in the traffic lights is effected by a switch in the motorman's cab.
"The track sanding gear for both forward and rear application is air operated from a valve pedal adjacent to the driver's seat, the supply of sand being fed automatically from storage hoppers under the staircases at each end.
"Life-guards of the County Council's standard gate and tray type give protection at the front and at both sides of each platform, as far back from the face of each collision fender as the radial movement of the trucks will permit, so that an effective time lag for definite operation of the pick-up tray is obtained.

Electrical Equipment

The complete power equipment consists of four motors, each nominally of 30/35 H.P., mounted one to each of the four axles, so that full 100 per cent, traction efficiency is obtained for rapid acceleration on the level with increased speed and stability for mounting heavy gradients. Each of the two truck pairs of motors is permanently electrically coupled in series, so that for purposes of control each truck pair is operated exactly as a single motor on an ordinary maximum traction two-motor type of equipment. This arrangement permits of acceleration from rest at a rate of 3-5 ft per second up to a free running speed of 30-35 miles per hour on the level, with corresponding speed of 12-15 miles per hour on an up gradient of 1 in 11. The drive is taken through single helical gears specially ground to ensure silent running and mounted on shafts running in roller bearings, with gear cases of non-drumming type, containing a specially selected anti-friction compound.
"The controllers are of the horizontal cam-operated type, arranged for crossed field braking, to provide for magnetic braking in either a forward or backward direction of travel in relation to the driving end of the vehicle, without need for the motorman to move the reverse key.

Brakes

The car is equipped with wheel and track brakes. All wheel brake blocks are arranged for either air or hand application, and the track brakes by air and magnetic application. The dual control valve for the air brake application to the wheel and track brake is on the top of the controller and the operating mechanism is interlocked with the magnetic brakes and power. Manual control of the air wheel brake is effected by means of a separate handle which is interlocked with the main controller handle for automatic quick release immediately prior to the application of power. The air application to the track brake is controlled by a device coupled to the main controller handle which comes into operation on the first stop from the off position on the brake side and prior to the magnetic application.

"A special device provides for the air application to the wheel brakes being automatically released immediately prior to application of magnetic braking. Automatic quick release valves close to the brake cylinders, which come into operation on a quick exhaust of air through the motorman's valve, ensure rapid exhausting of the cylinders to provide a safeguard from driving against the air brakes with its consequent waste of power.

"The air compressor is electrically driven direct from the line current, and is automatically controlled. It gives at normal voltage an output of 12 cubic feet of free air per minute, storage for which is provided in five cylindrical welded steel reservoirs having a nominal capacity of 6-5 cu. ft. Three of these (together with the special unit separately reserved for operation of the doors and steps) are accommodated in the cab at one end, with the remaining unit at the other end of the car. The storage capacity referred to meets amply all requirements in regard to operation of brakes, doors and steps, track sanders, cab-screen wipers, with a sufficient margin for air operation of the main changeover (trolley and/or conduit operation), power and lighting switches, when needed at the junctions between routes equipped respectively with conduit and overhead trolley systems of traction.

Trolley Poles

Further to simplify and speed up the change-over from either trolley or conduit operation to the other, the car is furnished with two separate overhead trolleys, so avoiding the need for the boom to be swung in traffic.

Trucks

The trucks are constructed of standard rolled steel sections with shaped plate side frames, of convenient depth and length to accommodate two motors, with space for magnetic track brake shoes, and accommodation for the application of independent mechanical or air-operating slipper brake devices thereto within a wheel-base of 4 ft 9 in. for each unit. The axles are specially prepared to suit the motor gears and cases referred to with special attention to all practicable elimination of vibration or drumming. Each journal is ground to gauge to suit the specially made roller bearings, which are incorporated in journal boxes fitted with radial arms. Thus contact or wear as between lateral or transverse faces of either the track or journal boxes is eliminated, and all traction loads or lateral thrusts are conveyed from the axles through a series of king pins fitted with readily adjustable and renewable bushes direct to the main side frames of the truck, which are located to conform, as nearly as practicable, with the corresponding running centre of each motor armature shaft. The effect of this as regards elimination of noise and need for frequent adjustment of wearing faces is particularly marked.

"The laminated axle-box springs are of exceptional length and elasticity, and in addition are cushioned at each end by auxiliary coiled springs with cushioning washers, the whole arrangement closely following that used on the very latest types of high-speed railway rolling stock, whereby road and crossover shocks are substantially damped, with considerable advantage to the maintenance of equipment, and to the comfort of passengers. The side swinging bolsters of the trucks are fitted directly under the body fulcrum pins, so that the bearing load is carried continuously in a plane parallel with the axles and square to the track, so avoiding any transmission of diagonal road shocks to the body structure and consequential discomfort to passengers. The bolster springs are arranged in horizontal nests directly under the centre of each bolster, and provided with conveniently adjustable tension rods, whereby the permissible rise and fall of the bolster is governed through a balanced series of bell cranks and levers.

"The trucks throughout are built on sound production lines with fitting bolts and pins for all structural purposes, amplified by registering edges on removable parts or details, so that the possibility of noise or chatter arising therefrom has been reduced to the fullest practicable extent. For a similar purpose and to maintain constant definite relationship of truck centres, the bolster pivotal bolts are of unusually sturdy dimensions and of such formation as permits the utilisation of ball bearings for their accommodation and accurate location."

Appendix 2
Hidden Stories
Hannah Bale

During the deconstruction of LCC No. 1, the Conservation Workshop's staff and volunteers found themselves becoming archaeologists, as a large number of objects were brought to light, some of which had remained unseen for the best part of 80 years. They were found in light fittings, under the stairs and behind the internal lining panels. All had been dropped, or perhaps purposely inserted by passengers and the conductors who worked on the tramcar. None of them are large, but they give a glimpse into the lives of the people who handled them and an image of the times they lived in.

South platform cab door with the ventilation panel pulled back, revealing the artefacts inside; and one of the pennies found amongst the rust and dust. (National Tramway Museum)

Some of the items relate directly to the tramcar, its building, maintenance and service. Some could have been dropped or lost at any time during the tramcar's working life, but others relate directly to its time in either London or Leeds. All the items discovered during the deconstruction of LCC No. 1 were very dirty, as would be expected. For some 80 years the tramcar's less accessible spaces have been accumulating dust and grime, and naturally this has covered the hidden items. Some of it may be soot caused by the fires which occurred under the staircases and the centre of the tramcar while it was in London, and the electrical cable fire which occurred during its service in Leeds.

A heavily corroded item appears to be a rule. It was found behind one of the internal lining panels. One end is semicircular and has a hole through the metal, which looks to be purposeful rather than produced over time with wear. It could have been used to hang the rule on a nail in a workshop.

Three paper money bags were found, each one intended to contain five shillings' worth of copper, indicated by the printing on the bags. These were issued by the London Transport Executive and London Passenger Transport Board. The money they contained might have been used by the conductors as a 'float', to be given as change when issuing tickets. It was more usual, however, for the conductor to place money in them when counting his takings after or part way through his shift. One of the money bags was purposely folded and made into a semicircle, perhaps by the conductor or driver while on a break or in a quiet moment during a shift.

Tickets discovered behind the wall panels were collected and bagged up before being cleaned and recorded. (National Tramway Museum)

The metal rule found behind the internal wall panelling. (National Tramway Museum)

One of the items found relating to the tramcar's service in Leeds is a Foreman's Request bill. This records the need for half a pound of ¾ inch strong panel pins, possibly for use on repairs. The slip was filled out on 11 March 1952, and sheds light onto the work carried out in the first year of the tramcar going to Leeds.

Relief Slips

Several Relief Slips were discovered which had been filled out by No. 1's conductors. Eight of these belong to the tramcar's time with Leeds City Transport. Two of the slips have the conductors' names. These are Mr Little, Conductor No. 2132, and Mr Auckley, No. 2689. The other

LPTB and LTE money bags, made to carry five shillings' worth of pennies and halfpennies. (National Tramway Museum)

six have only the conductors' numbers, 2671, 2166, 266 and 755. Three of the slips were filled out by Conductor 2166. On these slips the conductors recorded the number of the first ticket they issued when starting their shift, from child fares to the variously priced adult tickets, and the number of the last ticket they issued. The day of service was also recorded. Some conductors put down more details than others. There are slips on which only the day of the week or of the month, or both, were noted, but most helpfully for us, Conductor 755, Mr Little, and Conductor 266 put the exact dates, 17/3/1952, 22/11/1952, and 1/8/1954 respectively. Some of the slips also have the services the tramcar served on them, such as Moortown, West Park, or Moortown to Lawnswood. The back of the slips had forms to fill out if a ticket was cancelled or lost property was found. Below these are lines for the conductor's details and shift, allowing discrepancies and lost property to be traced back to the conductor who recorded them. These forms had not been filled out. Completed forms would have been handed in to the Tramway Office at the end of the shift.

The earliest Relief Slip was issued by the London Passenger Transport Board (Trams and Trolleybuses) and is a Relief Way-Bill (sic) slip. The conductor was Mr. Beech, No. 1441. He began that particular relief trip at 11.49am on 29/9/1938. The route was No. 16; he filled out the slip at Savoy Street and completed it on reaching Telford Avenue.

Other items found are from the tramcar itself, such as a fabric rope knot which was part of the chain or rope barrier to deter people from getting on or off the tram at that exit, before a sturdy metal bar was installed. Another is a painted metal bracket with a hoop which is thought to be one of the modifications made by Leeds City Transport. It is painted in the Leeds livery red. The bracket was attached to the outside of the tramcar, above the driver's window, for the bow rope, connected to the bow collector, to pass through the hoop allowing the bow collector to be raised or lowered.

Leeds City Transport Relief Slips, filled out by Mr Auckley (left) and Mr Little (right). These were the only Leeds slips found on which the conductors had signed their names. (National Tramway Museum)

It appears to have been removed at Crich when No. 1 was repainted in London Transport livery, and stored inside the tramcar until the deconstruction.

A few items are not easy to identify, such as a piece of frosted glass, with a red rectangle through part of it. There is evidence of it having broken twice, once quite recently as the broken section through the thickness of the glass is still clean. The other break is much older, and the broken area worn dull. There is no evidence to show where this glass came from, but the best idea is that it is part of a light covering from one of the saloons as the glass is frosted.

A particularly interesting and unexpected find was a quantity of sawdust behind the panel in the north end cab, near the controller. It is believed that the sawdust was purposely poured into the gap between the inner panel and outer body to act as insulation. Sawdust has sound dampening properties which would make it a good packing material for that panel, to lessen the noise of the compressor next to it.

Under the north end stairs was also found a thick, heavy woven canvas bag filled with sand. This would have been an essential piece of equipment, used to fill up the sandbox which is strategically placed to direct sand onto the rail head when the rails become slippery in icy or wet weather.

Tickets

The most numerous items found were tickets, found in layers behind the internal wall panels. Those on top were the most recent, issued by Leeds City Transport, dating to the 1950s. Below these were London Passenger Transport Board (LPTB) tickets, and, at the bottom, those issued by London County Council, dating to the 1930s and 1940s. The tickets are now very fragile, encrusted with decades of dirt, which usually obscures one side – the side facing upwards - making it unreadable. The side facing downwards was protected to an extent from the falling dust. Although it is hard to tell, most of the tickets appear to be one colour, a pale fawn, but some stand out as yellow, blue or pink. These brightly-coloured tickets were issued by Leeds City Transport and being more recent have sustained less discolouration, fading and encrusting with dirt than the older London issues.

Some tickets are torn in half, or were purposely screwed up, while others are complete; some are quite short, others are very long, depending on the city they were issued in. One of the longer tickets was issued by the LPTB and reads "Scholar (Under 18) Prepaid. Valid for any 2d adult direct ordinary single fare stage on the Board's tramcars or trolleybuses only. Not available on Sundays and School Holidays". Down the side of each long edge are days of the month and on the reverse are terms and conditions, although these are now too much obscured to read. One of the torn tickets found, issued by Leeds City Transport, cost 4d for one child.

Many of the tickets are of interest due to the advertisements they carry on the back, promoting a wide range of items for sale, companies, and entertainment venues.

Advertisements: health and nutrition

Ticket-holders are advised of Bovril's health-giving properties: "Bovril. Gives fitness without fatness" and "Keep fit on Bovril". One ticket even promotes safety on the trams, advising passengers not to jump from moving cars unless they want to drink their Bovril in hospital!

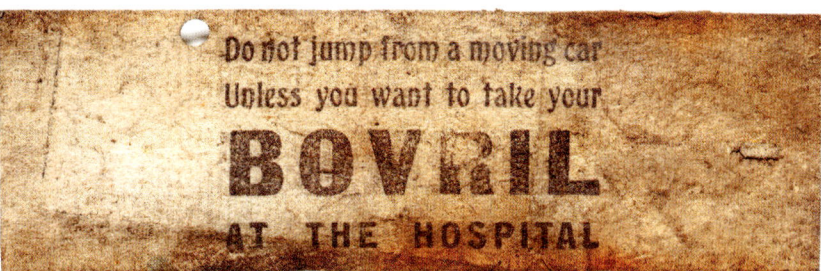

"Blairs Pills for Rheumatism & Gout" were available at "Boots and All Chemists". Mr. Blair (a pseudonym) was a pioneer of the production of muscle building protein supplements and, later, nutritional supplements [1].

"Liqufruta cure for everything", a natural garlic oil cough medicine, is still manufactured today in Devon [2], and "ENT Powders for Colds", selling at two shillings, promised to "cut through the most violent cold in a single night."

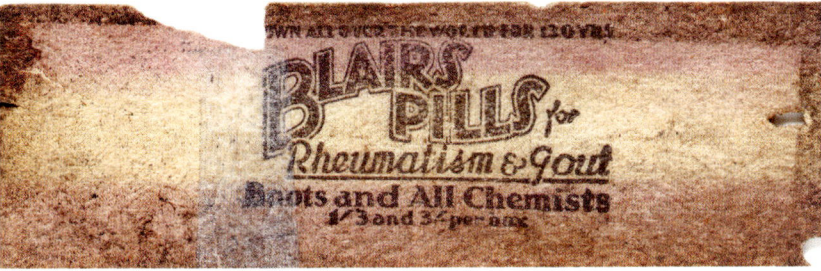

(National Tramway Museum)

Advertisements: food and drink

"You'll travel a long way before you find the equal to Hooper Struve's Mineral Water" one slogan claims. Hooper Struve still manufactures soft drinks [3].

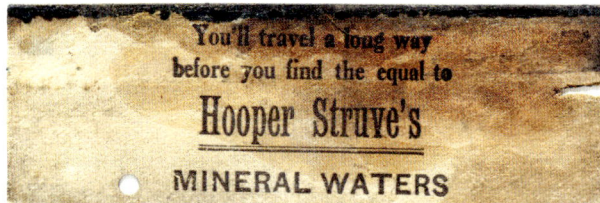

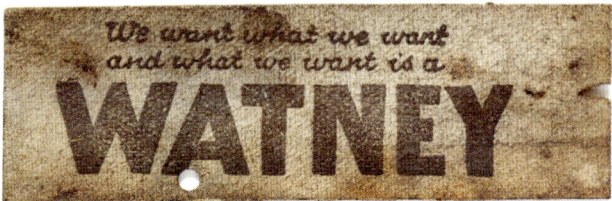

(National Tramway Museum)

Hayward's advertised their Military Pickle – "Still the Best". The Hayward Brothers established their company, based in Montford Place, Lambeth, in 1868, making pickles, vinegar, sauces and cordials, with Military Pickle being one of their specialities [4]. Hayward's is also still in existence.

Another household name is Camp Coffee, the advert stating "No Standing! Ready in a minute." Camp Coffee, created in 1876, was the world's first instant coffee [5].

There are several advertisements for Watney's Beer and Reid's Stout. Founded in 1837, Watney's was one of London's foremost breweries, producing the first keg beer in Britain. After the company closed in the 1970s, its 'Red Barrel' brand has recently been resurrected and is once again on sale [6]. Reid's was the only stout brewed in England [7], and the advert states it is "to shout about". It continued to be brewed until 1961 [8].

Advertisements: home products

Examples include "Disney's Furniture from factory to your door" and Marconiphone. The Marconiphone was manufactured by Marconi's Wireless Telegraph Company, Chelmsford, the first wireless factory in the world [9]. Marconiphone is stated to be "The FIRST and FINEST of all RADIO". By this time the company had moved to Tottenham Court Road, the address given on the advert to write to, to receive their company catalogue "A City of Sound".

(National Tramway Museum)

Advertisements: businesses

Some tickets carry advertisements for businesses such as Morley's of Brixton [10], which opened in 1880 and is still Brixton's top department store [11].

(National Tramway Museum)

Harringay Stadium and Arena in South-West Tottenham, in its heyday, hosted as many major sporting events as Wembley [12]. The Arena is promoted as a place to watch and play ice hockey and go ice skating, and Harringay Park as a venue for greyhound racing. Other venues advertised include the

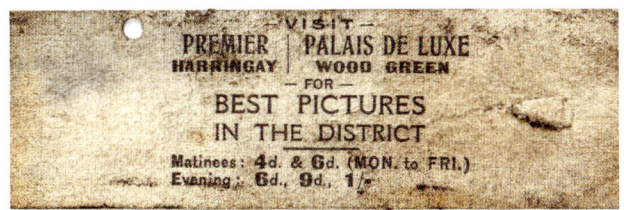 

(National Tramway Museum)

Premier in Harringay and the Palais de Luxe in Wood Green, the best cinemas in the district, with "Matinees: 4d & 6d (Mon. to Fri.) Evening: 6d, 9d, 1/-".

There are two adverts for "Renno's of 232-3-4 Upper Street Islington and 27 Stockwell Street Greenwich". One of these carries an advert for "cycles 2/- weekly Largest stock in London", the other promotes motorcycles, particularly the renowned British Montgomery. Other tickets show adverts for Farringdon Friendly Collecting Society; The London Assurance, an insurance company still providing cover today [13]; Leicester Permanent Building Society, now part of Santander [14]; "Scrubb's Marina Bath Cubes"; N. Berg "The Taylor (sic) of to-day", with shops at 127/8 Caledonian Road and 20 High Street and "'Angel' trams [passing] the doors"; and the RSPCA "Guardians of the nation's animals", asking for donations.

Advertisements: Leeds City Transport tickets

The majority of the tickets issued by Leeds City Transport do not carry any advertisements on the reverse. However, some items do appear, such as "Melbex" (Melba Toast), which would help the consumer "Slim and keep fit"; Vantella, a brand of men's shirt, extolling the longevity of the collar and cuffs; and a type of knitting machine which, it is claimed, will enable the purchaser to "Knit 50 times faster". The small illustration shows a ball of wool attached to a machine which appears to make knitted fabric, to be cut and sewn into a garment.

(National Tramway Museum)

Most interesting, however, are the two adverts connected to Yorkshire. The first promotes "Woods Genuine Pork Pies", as manufactured by Wood's Bacon Factory Limited, Mirfield, Yorkshire. Situated just over 10 miles from Leeds, the factory was founded in 1938 under Mary Wood, Company Director [15]. Pigs were kept in pens at the side of the factory before being taken inside for slaughter [16]. The Factory also had an onsite bakery where the pork pies were made [17].

The second advert relates directly to Leeds, advertising excursions to "the sunny coast and lovely countryside" offered by Wallace Arnold, situated at that time at 59 Corn Exchange, Leeds. Founded early in the twentieth century, until 2019 the company provided day trips to the Yorkshire coast for people living in the city [18].

Coins

A few coins were found. Whoever dropped them, the loss could have been quite significant; a less wealthy passenger could greatly regret losing even a penny, while a conductor would have had to account for the discrepancy in their takings. The coins are: a 1908 penny; two 1918 pennies; a 1919 penny; two halfpennies, one dated to 1913 and the other to 1924; and a 1930 silver sixpence. The latter was found in the north platform light fitting in the ceiling and fell out when the light was removed. Although this coin may have fallen into the light from the upper deck through a hole, it also might have been purposely inserted into the light during building as a good luck charm, as the belief in 6d coins as a lucky charm was still prevalent in the 1930s [19].

Cigarette packets and a cigarette card

Two cigarette packets reflect the time when smoking on public transport was still permissible. One packet is torn; the brand is Senior Service, made by J. A. Pattreiouex Ltd, Manchester. On one side is a crest with an anchor and olive or laurel leaves surmounted by a crown; on the other side are two seagulls, a ship surrounded by laurel leaves, and again a crown at the top. The packet was probably deliberately torn, as on the inside are written some letters and figures. Perhaps one of the conductors used the packet to work out the fares and takings at the end of a shift, before writing them up on the Relief Slip. The second is a complete Navy Cut Tobacco cigarette packet made at the Castle Tobacco Factory, Nottingham by John Player & Sons – the famous "Player's" cigarettes which are still

The Senior Service cigarette packet. Letters and figures have been written on the reverse side.
(National Tramway Museum)

The Player's cigarette packet, which was found together with the waybill slips. Unlike the Senior Service packet, it was complete. (National Tramway Museum)

manufactured today in Nottingham [20]. On one side of the packet is depicted a seascape and sailor and on the other a small picture of Nottingham Castle with the company lettering and motto. The packet originally contained 10 medium cigarettes.

As well as the cigarette packets, a cigarette card was found from a packet of Wills's Cigarettes, showing Stanford Robinson, in a series of Radio Celebrities. Born in Leeds in 1904, by the 1930s Stanford Robinson had become one of the most prominent conductors, arrangers and recorders of music [21].

An envelope for donations

A more unusual find was an envelope to take donations for 'The Priests (sic) Training Fund (Leeds Ecclesiastical Education Fund)'. The instructions on the envelope state that the donations are to be given to the donor's Parish Priest or placed in the collection on the first or second Sunday of Advent. Although we can't know the year, the time limit stated on the envelope gives quite a precise time frame for when the envelope was probably dropped, perhaps the middle or end of November to the first couple of weeks of December.

(National Tramway Museum)

Confectionery wrappers

These include the papers from a piece of Lyons' Chewing Gum. The inner paper is pink or orange and the outer paper yellow with orange and dark pink or red writing and white edging. There is also a box which contained Welch's Luxury Toffee, produced by Welch & Sons Ltd, Tynemouth. The box is very plain, with no ingredients list or nutritional information. Only the front has any real detail on it, with a dark brown section giving the company name and product, and the other half showing squares of toffee.

The Lyons' Chewing Gum and Welch's Luxury Toffee packets, after they were cleaned. (National Tramway Museum)

A Christmas card

Behind one of the panels of the lower saloon, beneath one of the windows, a Christmas card was found. It is single sided, like a postcard, and has no festive picture to decorate it, only a border of holly leaves and berries. The text, now partly obscured with dust, reads: "The Mayor and Mayoress of Islington (Councillor G. B. Naish, J. P., L. C. C., and Mrs. Naish), wish you a Merry Christmas and a Happy New Year." The date is 24 December, so was given the day the council staff broke up for their Christmas holidays. The top of the card also has the Coat of Arms of Islington. The text and border are printed, so it is probably one of a print run of cards given to the Mayor's staff or staff of Islington's local council.

George Bryant Naish, born in Bethnal Green in 1874 [22], was a member of the Labour Party and a politician in Islington, becoming mayor of Islington in 1936-1937. He was also a member of London County Council as Islington West representative from 1928 to 1949 [23].

The Christmas card when it was found on removing the internal panelling, amongst the other artefacts and debris. (National Tramway Museum)

Finally, two of the most personal and poignant items to be found were two First World War medal ribbons and a comb.

The medal ribbons

The two medal ribbons were discovered amongst some of the discarded tickets, so covered in soot and dust that it was not until the tickets were being conservation cleaned that the ribbons were identified. Made of watered silk, the first is the ribbon of The British War Medal 1914-1918. Its two bands of royal blue and black are still visible at the outer edges of the ribbon, although the black appears more as purple. The bands of white and the wide central band of orange have now faded into each other and been obscured by dust; however, a small section of white can still be made out on the reverse. The second ribbon, also made of watered silk, is the double rainbow ribbon of The Victory Medal,

First World War ribbons, known as 'Mutt and Jeff', found amongst the dropped tickets. (National Tramway Museum)

with red in the centre radiating outwards through the spectrum to violet at both sides [24]. The ribbons are folded into narrow strips, with a support inside to keep them stiff and maintain their shape, sewn closed with white thread and with a single stitch to anchor them together. In the 1920s, the medals had been given affectionate nicknames, together becoming known as 'Mutt and Jeff', taken from the names of characters in a popular comic strip [25].

We will never know who the ribbons belonged to. Men who wore a uniform as part of their employment, including tramway drivers and conductors, were permitted to wear their ribbons sewn onto their tunics. Although less likely, it may even be possible the ribbons belonged to a lady, as women who had served were eligible to be awarded the medals [26].

The comb

The comb was also found amongst the layer of tickets that had fallen into the wall panelling. It was so caked with dust and dirt that at first it appeared black, but when an area was rubbed clean, the comb was seen to be made of a creamy-yellow, almost opalescent coloured plastic. This is very typical of the time, with plastics manufactured to resemble luxury materials, such as mother of pearl, tortoiseshell and ivory. On the end of one side of the comb is printed "Foreign", indicating that it was made overseas and imported into Britain before sale.

As with all the items found, we can only speculate about who the original owner was, but the comb is the most personal and relatable item to be found during the deconstruction of the tramcar.

The items discovered during the deconstruction of LCC No. 1 have been carefully cleaned and stored in protective boxes for preservation into the future. They all help tell not only the story of the tramcar, but also of the decades it was in service and the people who worked and travelled on it.

The comb, with its 'Foreign' stamp to the right-hand side, after it was cleaned. (National Tramway Museum)

References

Chapter 1
[1] LCC Highways Committee LCCTT doc.
[2] *Sunday Times*, 1932, 8 May.
[3] Tremayne, D, 2005, *Donald Campbell: The Man Behind the Mask,* Bantam Books.
[4] Elliott, R, 1989, *Letter to TLRS Journal*, in *Tramfare* No. 127.
[5] Price, J H, 1958, *LCC Number One,* in *Modern Tramway* Vol. 21, p. 154.
[6] Reed, J, 1997, *London Tramways*, Capital Transport Publishing, p. 76.
[7] Oakley, E R, 1991, *London County Council Tramways Vol. 2*, LTHG, p. 752.
[8] Crewe, T, 2016, *The Strange Death of Municipal England* in *London Review of Books*, Vol. 38 No. 24.
[9] Hall, C, 1977, *Sheffield Transport*, The Transport Publishing Company, pp. 201-7.
[10] Soper, J, 2003, *Leeds Transport, Vol. 3 1932-1953*, Leeds Transport Historical Society, pp. 953-74.
[11] ibid., p. 961.
[12] Staddon, S A, 1964, *The Tramways of Sunderland*, The Advertiser Press Ltd., pp. 74-6.
[13] Horne, J B, and Maund, T B, 1987, *Liverpool Transport Vol. 3 – 1931-9*, LRTA, pp. 90-134.
[14] Stewart, I, 1983, *The Glasgow Tramcar*, Scottish Tramway Museum Society, pp. 100-5.
[15] Buckley, R, 1975, *History of Tramways, from Horse to Rapid Transit,* David and Charles, pp. 140-4.
[16] Waller, P, 1994, T*he Heyday of the European Tram*, Ian Allan Publishing, p. 59.
[17] Buckley, R, 1975, *History of Tramways, from Horse to Rapid Transit*, David and Charles, pp. 140-4.
[18] Hearse, G S, 1965, *The Tramways of Gateshead*, George S Hearse, p. 76.
[19] Smeeton, C S, 1986, *The Metropolitan Electric Tramways, Volume 2 – 1921 to 1933*, Light Rail Transit Association, p. 399.
[20] Price, J H, 1993, *Paper 3: London Tramway Technology* in Higginson, M (ed.), *Tramway London: Background to the Abandonment of London's Trams 1931 – 1952*, LRTA, p. 44.
[21] Oakley, E R, 1991, *London County Council Tramways Vol. 2*, LTHG, pp. 728-33.
[22] ibid., pp. 740-8.
[23] ibid., pp. 854-7.

Chapter 2
[1] Inwood, S, 2005, *City of Cities: the Birth of Modern London*, Pan Macmillan, p. 9.
[2] Collins, P, 2001, *London's Trams: A View from the Past*, Ian Allan Publishing, p. 47.
[3] Local Government Act, 1888.
[4] Collins, P, 2001, *London's Trams: A View from the Past,* Ian Allan Publishing, p. 47.
[5] Inwood, S, 2005, *City of Cities: the Birth of Modern London,* Pan Macmillan, p. xvii.
[6] Yearsley, I, 1993, *Paper 2: Previously Unexplored Aspects of London's Tramway Finances*, in Higginson, M (ed.), *Tramway London: Background to the Abandonment of London's Trams 1931 – 1952,* LRTA, p. 17.
[7] Green, O, and Higginson, M, 1993, *Paper 1: London's Tramways in their Years of Decline*, in Higginson, M (ed.), *Tramway London: Background to the Abandonment of London's Trams 1931 – 1952*, LRTA, p. 6.
[8] Oakley, E R, 1991, *London County Council Tramways Vol.2 North London*, London Tramways History Group: Kent in association with the TLRS and LRTA, p. 752.
[9] Williams, D, 1963, *London and the Financial Crisis in The Economic History Review* New Series, Vol. 15, No. 3, pp. 524-6.
[10] Yearsley, I, 1993, *Paper 2: Previously Unexplored Aspects of London's Tramway Finances*, in Higginson, M (ed.), *Tramway London: Background to the Abandonment of London's Trams 1931 – 1952*, LRTA, p. 23.
[11] UK: http://www.bankofengland.co.uk/statistics/rates/baserate.xls, 08.04.10; US: http://www.nber.org/databases/macrohistory/contents/chapter13.html, 08.04.10. Cited in Middleton, R, 2010, *Oxford Review of Economic Policy,* Volume 26, Number 3, p. 423.
[12] Tramways Act 1870, section 20.
[13] Yearsley, I, 1993, *Paper 2: Previously Unexplored Aspects of London's Tramway Finances*, in Higginson, M (ed.), *Tramway London: Background to the Abandonment of London's Trams 1931 – 1952*, LRTA, p. 25.
[14] Smeeton, C S, 2000, *The London United Tramways, Vol. 2 1913-33*, LRTA, p. 404.
[15] Smeeton, C S, 1986, *The Metropolitan Electric Tramways, Volume 2 - 1921 to 1933*, LRTA, p. 399.
[16] Green, O, and Higginson, M, 1993, *Paper 1: London's Tramways in their Years of Decline*, in Higginson, M (ed.), *Tramway London: Background to the Abandonment of London's Trams 1931 – 1952,* LRTA, p. 8.
[17] Yuzawa, T, 2014, ch. 6: *Urbanisation and Transport Restructuring before World War II: A Comparison Between London and Osaka,* in Moraglio, M and Kopper, C, *The Organization of Transport: A History of Users, Industry, and Public Policy,* Routledge, p. 96.

[18] ibid., pp 96-7.
[19] Oakley, E R, 1991, *London County Council Tramways Vol.2 North London*, London Tramways History Group: Kent in association with the TLRS and LRTA, p. 504.
[20] Royal Commission on Transport, 1931, *Final Report 1930-1*, (Cmd.3751), para. 368.
[21] Collins, P, 2001, *London's Trams: A View from the Past,* Ian Allan Publishing, p. 91.

Chapter 3
[1] *Tramways Act 1870*, Section 28*.
[2] ibid., Section 10.
[3] ibid., Section 43.
[4] ibid., Section 5.
[5] ibid., Section 9.
[6] Barker, T C, and Robbins, M, 1963, *A History of London Transport Vol. 1 – The Nineteenth Century,* Allen & Unwin, pp. 179-80.
[7] *The Times*, 1958, 16 March, cited by Barker, T C, & Robbins, M, 1963, *A History of London Transport Vol. 1 – The Nineteenth Century*, Allen & Unwin, pp. 90-1.
[8] Collins, P, 1995, *The Tram Book*, Ian Allan Publishing, p. 7.
[9] Lee, C E, 1953, T*he English Street Tramways of George Francis Train – II, The Journal of Transport History,* pp. 97-108. See also *Grace's Guide to British industrial history*, available at: https://www.gracesguide.co.uk/George_Francis_Train
[10] Voice, D, 2010, *George Francis Train*, in *Tramfare,* May 2010. Also available at: https://www.tramwayinfo.com/Tramframe.htm?https://www.tramwayinfo.com/tramways/Articles/Train.htm
[11] Barker, T C, and Robbins, M, 1963, *A History of London Transport Vol. 1 – the Nineteenth Century,* Allen & Unwin, p. 181.
[12] Royal Commission on London Traffic, 1905, *Report of Royal Commission appointed to enquire into the means of locomotion and transport in London*, Cd. 2597, paras. 114 & 115.
[13] Price, J H, 1993, *Paper 3*: *London Tramway Technology,* in Higginson, M (ed), *Tramway London: Background to the Abandonment of London's Trams 1931 – 1952*, LRTA, p.48.
[14] Wilson, F E, 1961, *The British Tram*, Percival Marshall, p. 74.
[15] Price, J H, 1993, *Paper 3*: *London Tramway Technology,* in Higginson, M (ed), *Tramway London: Background to the Abandonment of London's Trams 1931 – 1952*, LRTA, p.48.
[16] ibid., p. 49.
[17] Royal Commission on London Traffic, 1905, *Report of Royal Commission appointed to enquire into the means of locomotion and transport in London,* Cd. 2597, para. 92.
[18] ibid., para. 95.
[19] Robson, W A, 2013, *The Government and Misgovernment of London,* Routledge, p. 142.
[20] *Report of the Select Committee on Metropolitan Communications (1855)*, B.P.P., 1854-5, vol. x, p. iv.
[21] Barker, T C, and Robbins, M, 1963, *A History of London Transport Vol. 1 – the Nineteenth Century,* Allen & Unwin, p. 191-2.
[22] ibid., p. 85.
[23] Robson, W A, 2013, *The Government and Misgovernment of London,* Routledge, p. 145.
[24] Royal Commission on London Traffic, 1905, *Report of Royal Commission appointed to enquire into the means of locomotion and transport in London*, Cd. 2597, para. 112.
[25] ibid., para. 115.
[26] ibid., para. 201.
[27] Barker, T C, and Robbins, M, 1963, *A History of London Transport Vol. 1 – the Nineteenth Century,* Allen & Unwin, p. 210.
[28] *Electric Railway Journal, 1924*, Vol. 64, no. 12, 20 September, pp. 425-47.
[29] Barker, T C, and Robbins, M, 1963, *A History of London Transport Vol. 1 – the Nineteenth Century,* Allen & Unwin, p. 196.
[30] ibid., p. 263.
[31] Transport for London Corporate Archives Research Guides *Research Guide No 14*.
[32] Barker, T C, and Robbins, M, 1974, *A History of London Transport Vol. 2 – the Twentieth Century to 1970*, Allen & Unwin, p. 122.
[33] Robson, W A, 2013, *The Government and Misgovernment of London*, Routledge, p. 148.
[34] Ball, M, and Sunderland, D T, 2002, *An Economic History of London, 1800 – 1914*, Routledge, pp. 258-62.
[35] Barker, T C, and Robbins, M, 1974, *A History of London Transport Vol. 2 – the Twentieth Century to 1970*, Allen & Unwin, pp. 164-6.
[36] ibid., pp. 166-70.
[37] ibid., pp. 172-3.
[38] ibid., pp. 77-84.
[39] Inwood, S, 2005, *City of Cities: The Birth of Modern London,* Pan Macmillan, p. 261.

[40] Catford, N, and Bolger, P, 1999, *Liverpool Overhead Railway*. Available at: https://www.subbrit.org.uk/sites/liverpool-overhead-railway/(Accessed: 28 September 2021).
[41] Ball, M, and Sunderland, D T, 2002, *An Economic History of London, 1800 – 1914*, Routledge, pp. 262.
[42] Inwood, S, 2005, *City of Cities: The Birth of Modern London,* Pan Macmillan, p. 247.
[43] Hansard, HL Deb11 Dec 12 vol 13 cc 172.
[44] Inwood, S, 2005, City of Cities: *The Birth of Modern London,* Pan Macmillan, p. 247.
[45] ibid., p. 248.
[46] ibid., p. 247.
[47] Ball, M, and Sunderland, D T, 2002, *An Economic History of London, 1800 – 1914*, Routledge, pp. 252.
[48] Inwood, S, 2005, *City of Cities: The Birth of Modern London,* Pan Macmillan, p. 249.
[49] ibid., p. 250.
[50] ibid., pp. 248-9.
[51] Joint Select Committee of the House of Lords and the House of Commons (1901), *Report on London Underground Railways, 23 July 1901*, HMSO: London, 1901, p. ix.
[52] Moraglio, M, and Kopper, C (eds), 2014, *The Organization of Transport: A History of Users, Industry, and Public Policy,* Routledge, p.93.
[53] Barker, T C, and Robbins, M, 1963, *A History of London Transport Vol. 1 – the Nineteenth Century*, Allen & Unwin, p. 189.

*This section remained in force until the Tramways Act was repealed in 1992.

Chapter 4
[1] Ministry of Transport, 1923, *Tramways and Light Railways (Street and Road) and Trackless Trolley undertakings returns 1919-1923*, HMSO.
[2] Ministry of Transport, 1920, letter to LCC, with certificate, 27 November.
[3] LCC Highways Committee Report, 1920, 9 December, *Tramway and railless proposals – session 1922.*
[4] *Tramways Act 1870*, Section 10.
[5] Hansard HL, Deb 11-Dec-12, vol 13, cc 172.
[6] Dunbar, C S, 1967, *Idealism and competition*, LRTL, p. 9.
[7] Royal Commission on London Traffic, 1905, Rep 26, June, Cd 2597.xxx.533   114.
[8] Dunbar, C S, 1967, *Idealism and competition*, LRTL, p. 3.
[9] Hansard Rep, 1905, 26 June, Cd 2597.xxx.533   111.
[10] LCC Highways Committee Minutes, 1920, 23 March, 22 April, 29 April, 7 May.
[11] Oakley, E R, 1991, *London County Council Tramways Vol 2,* LTHG, p. 713.
[12] Morris, O J, 1953, *Fares Please*, Ian Allan, p. 67.
[13] *Tramway & Railway World,* 1929, February 14, pp. 74-5.
[14] LCC Highways Committee Minutes, 1923, May.
[15] Oakley, E R, 1989, *London County Council Tramways Vol 1,* LTHG, p. 333.
[16] Barker, T C, and Robbins, M, 1974, *A History of London Transport Vol 2,* Allen & Unwin, p. 186.
[17] Brown, D, 2009, *Southern Electric Vol 1*, Capital Transport Publishing, pp. 11, 24.
[18] Tramway and Light Railways (Street and Road) and Trackless Trolley undertakings returns, 1922-1925.
[19] Dunbar, C S, 1967, *Idealism and competition*, LRTL, p. 9.
[20] LCC, 1921, *Services rendered to London by the LCC tramways,* LCC, November.
[21] Green, O, and Higginson, M, 1993, *Paper 1: London's Tramways in their Years of Decline*, in Higginson, M (ed.), *Tramway London: Background to the Abandonment of London's Trams 1931-1952,* LRTA, p. 6.
[22] ibid., p. 57.
[23] ibid., p. 6.
[24] Barker, T C, and Robbins, M, 1974, *A History of London Transport Vol 2*, Allen & Unwin, p. 223.
[25] ibid., p. 224.
[26] *The Electric Railway, Bus and Tram Journal*, 1923, May 11.
[27] *The Electric Railway, Bus and Tram Journal*, 1923, June 22.
[28] Townsin, A A, 1980, *Blue Triangle*, TPC, p. 68.
[29] LCC Highways Committee, LCC Report 1924, 15 July.
[30] *Tramways and Light Railways Association Journal*, 1925, pp. 3144-5.
[31] Klapper, C, 1962, *The Golden Age of Tramways*, p. 113.
[32] LCC Highways Committee Minutes, 1920/1.
[33] LCC Highways Committee Report, 26 January 1921, *Assessors Report on competition for designs for a new type of tramcar.*
[34] ibid.
[35] *Tramways and Light Railways Association Journal*, 1923 conference, June.

[36] *Tramways and Light Railways Association Journal,* 1926 conference, June.
[37] *Tramways and Light Railways Association Journal,* 1926, *Report of Car Bodies sub-committee; Report of truck and Equipment sub-committee* pp. 3396-3436.
[38] ibid., pp. 3438-9.
[39] ibid., p. 3444.
[40] *The pros and cons of the trolley bus*, 1926, in *Commercial Motor*, 15 June.
[41] Tramways and Light Railways Association Journal, 1926 conference, June.
[42] Bishop, R A, 1931, *The Electric Trolley Bus*, Sir Isaac Pitman & Sons, Ltd, p. 83.
[43] ibid., p. 88.
[44] Smeeton, C, 1986, *The London United Tramways Vol 2*, LRTA, p. 424.
[45] Townsin, A A, 1980, *Blue Triangle*, TPC, p. 33.
[46] Smeeton, C, 1986, *The London United Tramways Vol 2,* LRTA, p. 425.
[47] ibid., p. 429.
[48] Blacker, K C, 1962, *The Felthams*, Dryhurst, p. 43.
[49] Harley, R, 2002, *LCC Electric Tramways*, Capital, p. 136.
[50] Oakley, E R, 1991, *London County Council Tramways Vol 2*, LTHG, p. 726.
[51] ibid.
[52] LCC Highways Committee Minutes, 1927, 12 October.
[53] Oakley, E R, 1991, *London County Council Tramways Vol 2*, LTHG, p. 727.
[54] ibid., p. 728.
[55] ibid.
[56] ibid.
[57] LCC Highways Committee Minutes, 1932, March.
[58] Riddell, J, 2010, *Tramway Art*, Capital Transport.
[59] Smeeton, C, 1986, *The Metropolitan Electric Tramways Vol 2*, LRTA, p. 366.
[60] Blacker, K C, 1962, *The Felthams*, Dryhurst, p. 14.
[61] *Tramways and Light Railways Association Journal*, 1926, Chair on rolling stock committee 1926.
[62] Smeeton, C, 1986, *The Metropolitan Electric Tramways Vol 2*, LRTA, p. 366.
[63] ibid.
[64] *Tramway & Railway World*, 1929, March.
[65] Smeeton, C, 1986, *The Metropolitan Electric Tramways Vol 2,* LRTA, p. 385-391.
[66] ibid., p. 401.
[67] Blacker, K C, 1962, *The Felthams*, Dryhurst, p. 21.
[68] LCC Highways Committee Minutes, 1927, 20 December.
[69] Oakley, E R, 1991, *London County Council Tramways Vol 2*, LTHG, p. 740.
[70] *Tramway & Railway World*, 1929, April 18, pp. 203-205.
[71] Oakley, E R, 1991, *London County Council Tramways Vol 2*, LTHG, p. 716.
[72] LCC Highways Committee Minutes, 1928, August.
[73] LCC Highways Committee Minutes, 1929, July.
[74] LCC Highways Committee Minutes, 1929, 24 October.
[75] LCC Highways Committee Minutes, 1929, 12 November.
[76] Oakley, E R, 1991, *London County Council Tramways Vol 2*, LTHG, p, 750.
[77] LCC Highways Committee Minutes, 1931, 5 March.
[78] Price, J H, 1974, *The Brush Electrical Engineering Co Ltd*, TLRS, p. 24.
[79] *Tramway & Railway World*, 1927, 17 March, pp. 135-6.
[80] ibid., pp. 136-8.
[81] Oakley, E R, 1991, L*ondon County Council Tramways Vol 2*, LTHG, p. 741.
[82] Young, A D, 1984, *The EMB Company and its Tramway Products*, in *Modern Tramway*, November, p. 363.
[83] Oakley, E R, 1991, *London County Council Tramways Vol 2*, LTHG, p. 742
[84] Blacker, K C, 1962, *The Felthams,* Dryhurst, pp. 13-19.
[85] Lawson, P W, 1971, *Birmingham Corporation Tramway Rolling Stock Part 16*, in *Modern Tramway*, November, pp. 374-379.
[86] Lawson, P W, 1971, *Birmingham Corporation Tramway Rolling Stock Part 17*, in *Modern Tramway*, December, pp. 414-419.
[87] Hall, C, 1977, *Sheffield Transport,* The Transport Publishing Company, p. 204.
[88] *Tramway & Railway World*, 1932, 14 April.

Chapter 5
[1] LCC Highways Committee Minutes, 1929, 24 October.
[2] LCC Highways Committee Minutes, 1929, 12 November.
[3] Sinclair, G F, 1932, *Experiences With Modern Tramcars*, in *Tramway & Railway World*, 12 May, pp. 234-40.

[4] Price, J H, 1993, *Paper 3: London Tramway Technology*, in Higginson, M (ed.), *Tramway London: Background to the Abandonment of London's Trams 1931-1952*, LRTA, p. 38.
[5] Oakley, E R, 1991, *London County Council Tramways Vol 2*, LTHG, pp. 752, 820.
[6] LCCT Manager's Report, 1929, 24 October.
[7] Sinclair, G F, 1932, *Experiences with Modern Tramcars*, in *Tramway & Railway World*, 12 May, pp. 234-40.
[8] LCC Highways Committee Minutes, 1931, 26 March, p. 195.
[9] Young, A D, 1985, *EMB Trams*, LRTA, p. 54.
[10] LCC Highways Committee Minutes, 1930, 30 December.
[11] Price, J H, 1977, *Hurst Nelson Tramcars*, p. 36.
[12] LCC Highways Committee Minutes, 1931, 17 September.
[13] LCC Highways Committee General Manager Report, 1931, 17 December.
[14] LCC Highways Committee General Manager Report, 1931, 5 November.
[15] LCC Highways Committee, 1931, 3 December.
[16] LCC Highways Committee Reports, 1930-1932.
[17] LCC Highways Committee Quotation, 1932, February.
[18] LCC Highways Committee Quotation, 1932, 26 April.
[19] Young, A D, 1985, *EMB Trams*, LRTA, p. 65.
[20] Horne, J B, and Maund, T B, 1987, *Liverpool Transport Vol 3*, LRTA, p. 93.
[21] Oakley, E R, 1989, *London County Council Tramways Vol 1*, LTHG, p. 296.
[22] ibid., p. 298.
[23] Oakley, E R, 1991, *London County Council Tramways Vol 2*, LTHG, p. 751.
[24] ibid., p. 747.
[25] Elliot, R, 1989, *TLRS Journal 127*, January.
[26] Buckle, J, 2020, emails to Richard Sykes, April.
[27] *Tramway & Railway World*, 1932, 12 May, p. 228.
[28] LCC Highways Committee Report, 1930, 29 August.
[29] Young, A D, 1984, *EMB and its Products*, in *Modern Tramway*, November, p. 367.
[30] Oakley, E R, 1991, *London County Council Tramways Vol 2*, LTHG, p. 752.
[31] LCC Highways Committee Minutes, 1930, 30 December.
[32] Oakley, E R, 1991, *London County Council Tramways Vol 2*, LTHG, p. 755.
[33] *Tramway & Railway World*, 1931, 12 December.
[34] G D Peters, Catalogue.
[35] LCC Highways Committee Report, 1932, February.
[36] Press notices, *Daily Telegraph* and *Daily Sketch*, 1932, 5 May.
[37] *Sunday Times*, 1932, 8 May.
[38] *The Pullman Review,* 1933, LCC.
[39] *Tramway & Railway World*, 1932, 13 May.
[40] *Electric Railway Bus and Tram Journal*, 1932, 13 May, p. 177.
[41] *Electric Railway Bus and Tram Journal Current Topics*, 1932, 13 May, p. 176.
[42] *Omnibus Magazine*, 1932, May, p. 104.
[43] *Omnibus Magazine*, 1932, June, p. 124.
[44] Sinclair, G F, 1932, *Experiences with modern tramcars*, in *Tramway & Railway World*, 12 May, pp. 234-40.
[45] Price, J H, 1958, *LCC Number 1*, in *Modern Tramway*, July, pp. 152-3.
[46] *Electric Railway Bus and Tram Journal*, 1932, 13 May, p. 178.
[47] ibid., p. 180.
[48] ibid., p. 181.
[49] Price, J H, 1958, *LCC Number 1*, in *Modern Tramway*, July, p. 152.
[50] Horne, J B, and Maund, T B, 1987, *Liverpool Transport Vol 3*, LRTA, pp. 95-6.
[51] ibid., p. 93.
[52] LCCTT Collection drawings.
[53] Oakley, E R, 1998, *London Transport Tramways*, LTHG, pp. 139-40.
[54] *Tramway & Railway World*, 1932, September.
[55] Wiseman, R, 1994, *The Tramways of Bombay*, in *Tramway Review* 159, p. 270.
[56] ibid., p. 275.
[57] Soper, J, 2003, *Leeds Transport Vol 3*, Leeds Transport Historical Society, pp. 961-76.

Chapter 6
[1] *The Electric Railway, Bus and Tram Journal*, 1932, May 13, p. 183.
[2] ibid.
[3] *Omnibus Magazine*, 1932, June, p. 124.
[4] Blacker, K C, 1962, *The Felthams*: *The Story of the Union Construction Company*, Dryhurst Publications, p. 47

[5]   Barrie, J, 2006, *LCC Trams of the 1930s*, in *Tramway Review* No. 207, p. 268.
[6]   Oakley, E R, 1989, *London County Council Tramways Vol 1*, LTHG, p. 308.
[7]   *Omnibus Magazine*, 1932, May, p. 104.
[8]   Oakley, E R, 1991, *London County Council Tramways Vol 2*, LTHG, p. 756.
[9]   Morris, O J, 1953, *Fares Please*, Ian Allan, p.69.
[10] Barrie, J, 1969, *North London Tramways 1938 to 52*, LRTL, p. 9.
[11] Oakley, E R, 1989, *London County Council Tramways Vol 1*, LTHG, pp. 339, 412, 854.
[12] Collins, S G, 1977, *The Wheels Used To Talk To Us: a London tramwayman remembers*, ed. Terence Cooper, p. 99.
[13] Barrie, J, 1969, *North London Tramways 1938 to 52*, LRTL, p. 9.
[14] *The Electric Railway, Bus and Tram Journal, 1932*, May 20, p. 216c.
[15] Buckle, J, 2017, email to Robert Morris, 30 October.
[16] Barrie, J, 1969, *North London Tramways 1938 to 52*, LRTL, p. 9.
[17] Collins, S G, 1977, *The Wheels Used To Talk To Us: a London tramwayman remembers*, ed. Terence Cooper, p. 99.
[18] Oakley, E R, 1998, *London Transport Tramways*, LTHG, p. 31.
[19] Beardsell, D, 2016, *Tramway Engineering Timeline: London County Council Tramcar 1*, The National Tramway Museum, p. 4.
[20] Simpson, A, 2006, *London Transport No 1: A Chronology* by Andrew Simpson, in *Tramway Review* No. 207, p. 301.
[21] *London Passenger Transport Act 1933*, Section 3(4).
[22] Beardsell, D, 2016, T*ramway Engineering Timeline: London County Council Tramcar 1*, The National Tramway Museum, p. 5.
[23] ibid.
[24] Simpson, A, 2006, *London Transport No 1: A Chronology* by Andrew Simpson, in *Tramway Review* No. 207, p. 301.
[25] Barrie, J, 1969, *North London Tramways 1938 to 52*, LRTL, p. 9.
[26] Oakley, E R, 1998, *London Transport Tramway*s, LTHG, p. 111.
[27] ibid., p. 372.
[28] ibid., p. 92.
[29] Collins, S G, 1977, *The Wheels Used To Talk To Us: a London tramwayman remembers*, ed. Terence Cooper, p. 99.
[30] Oakley, E R, 1998, *London Transport Tramways*, LTHG, p. 381.
[31] ibid.
[32] Wiseman, R J, 2006, *LRTL Tramway Tours 1938-1946*, in *Tramway Review* No. 207, p. 280.
[33] Beardsell, D, 2016, *Tramway Engineering Timeline: London County Council Tramcar 1*, The National Tramway Museum, p. 5.
[34] ibid.
[35] Simpson, A, 2006, *London Transport No 1: A Chronology* by Andrew Simpson, in *Tramway Review* No. 207, p. 302.
[36] *London Transport Memorandum for Submission to Supplies Meeting* to be held on 24th May, 1951.
[37] *London Transport Schedule I & II – The Tramcars and Prices; Provisional Date of Delivery*.
[38] *London Transport Schedule*, 1950, 14 July.
[39] Chamberlain, Darryl, 2012, *Charlton: Where London's Last Trams Went to Die.* Available at: charltonchampion.co.uk (Accessed: 2 February 2020).
[40] *London Transport Memoranda for Supplies Meeting* held on 24 May, 1951.
[41] Shawcross, J, 2014, *Collections Management Attitude Statement London County Council Tramcar Number 1*, LCC/LIB/001, The National Tramway Museum, UK.
[42] Simpson, A, 2006, *London Transport No 1: A Chronology* by Andrew Simpson, in *Tramway Review* No. 207, p. 302.
[43] ibid.
[44] Collins, S G, 1977, *The Wheels Used To Talk To Us: a London tramwayman remember*s, ed. Terence Cooper, p. 99.

Chapter 7
[1]   Hilditch, G, 1985, unpublished letter to Bradshaw, 30 June, MATTM:2016.69, The National Tramway Museum.
[2]   Young, A D, 1974, *Leeds Trams*, 1932-1939, Light Railway Transport League, p. 68.
[3]   ibid., pp. 68-71.
[4]   ibid., p. 71.
[5]   *Leeds Transport Committee Minutes 1950*, Item 16, p 33.
[6]   *London Transport Memorandum for Submission to Supplies Meeting* to be held on 24th May, 1951.
[7]   Smith, E, 2017, letter to Robert Morris, 18 September.

[8] London Transport documents Schedule 1, *The Tramcars and Prices; Provisional Date of Delivery.*
[9] Buckle, J, 2017, email to Robert Morris, 30 October.
[10] Shawcross, J, 2014, *Collections Management Attitude Statement London County Council Tramcar Number 1,* LCC1/LIB/001, The National Tramway Museum, UK.
[11] ibid.
[12] Steward, T, 1979, *Flats, Snow and Ice*, in *Modern Tramway*, Vol. 42, No. 494, p. 52.
[13] Ross, I, 2017, *Bluebird: A Work in Progress*, in *Belvedere Echo*, Issue 49, p. 5.
[14] Crabtree, M, 2017, *LCC1 – The Trucks [Part 2]*, in *The Tramway Museum Society Journal* No. 238, p 26.
[15] *The Leeds Tramway News*, 1954, Issue 23, p. 2.
[16] Steward, T, 1979, *Flats, Snow and Ice*, in *Modern Tramway* Vol. 42, No. 494, p. 52.
[17] Soper, J, 2003, *Leeds Transport Three 1932-1953,* Leeds Transport Historical Society, p. 1039.
[18] Gilks, M, 2017, email to Robert Morris, 4 December.
[19] Steward, T, 1979, *Flats, Snow and Ice*, in *Modern Tramway* Vol. 42, No. 494, p. 52.
[20] Shawcross, J, 2014, *Collections Management Attitude Statement,* LCC1/LIB/001, The National Tramway Museum, UK.
[21] *The Leeds Tramway News*, 1953, Issue 12, p. 3.
[22] *The Leeds and District Transport News,* 1956, Vol. 3, No. 50, p. 12.
[23] Gilks, M, 2017, email to Robert Morris, 4 December, and Thornburn, C, 2017, Leeds 301 (unpublished notes), July.
[24] *The Leeds and District Transport News*, 1956, Vol. 3, No. 50, p. 13.
[25] Smith, E, 2017, letter to Robert Morris, 18 September.
[26] *The Leeds and District Transport News*, 1955, Vol. 2, No. 34, p. 2.
[27] Thornburn, C, 2007, *Leeds – The Preservation Story*, in *Tramway Review,* 212, p. 56.
[28] *The Leeds and District Transport News*, 1955, Vol. 2, No. 44, p. 7.
[29] *The Leeds and District Transport News*, 1956, Vol. 2, No. 48, p. 2.
[30] Thornburn, C, 2017, *Leeds 301* (unpublished notes), July.
[31] ibid.
[32] Hilditch, G, 1985, unpublished letter to Bradshaw, 30 June.
[33] Smith, E, 2017, letter to Robert Morris, 18 September.
[34] ibid.
[35] *Thornburn, C, 2017, unpublished notes, July.*
[36] *The Leeds and District Transport News*, 1956, Vol. 3, No. 55, p 55, and 1957, Vol. 3, No. 60, p. 87.
[37] *The Leeds and District Transport News*, 1957, Vol. 3, No. 64, p. 116.
[38] *The Leeds and District Transport News*, 1957, Vol. 3, No. 66, p. 132.
[39] Young, A D, 1974, *Leeds Trams, 1932-1939*, Light Railway Transport League, p. 95.
[40] ibid., pp. 94-96.
[41] Mather, Hon. Alderman B, 1975, *Looking Back on Leeds*, in *Modern Tramway* Vol. 38, No. 455, p. 368.
[42] ibid., p. 366.
[43] *The Leeds and District Transport News*, 1957, Vol. 3, No. 69, pp. 151-152.

Chapter 8
[1] Price, J H, 1970, *Consultative Panel for the Preservation of British Transport Relics*, pp. 2-3.
[2] Lambert, Mark, 2016, *The Designation and Display of British Railway Heritage in the Post-War Decades,* Doctoral Thesis, University of Nottingham, pp. 126-130.
[3] British Transport Commission, 1951, *The Preservation of Relics and Records, Report to the British Transport Commission*, pp. 7-8.
[4] ibid.
[5] ibid., p. 12.
[6] ibid., p. 16.
[7] Price, J H, 1970, *Consultative Panel for the Preservation of British Transport Relics*, pp. 3-4.
[8] British Transport Commission, 1951, *The Preservation of Relics and Records, Report to the British Transport Commission,* p. 13.
[9] ibid., p. 15.
[10] Price, J H, 1970, *Consultative Panel for the Preservation of British Transport Relics*, pp. 4-5.
[11] ibid., pp. 5-7.
[12] ibid., pp. 7-8.
[13] British Transport Commission, 1951, *The Preservation of Relics and Records, Report to the British Transport Commission,* p. 17.
[14] ibid., p. 8.
[15] Price, J H, 1963, *News from the Museums*, in *Modern Tramway and Light Railway Review*, Vol. 26, No. 307, p. 237.

[16] Price, J H, 1970, *Consultative Panel for the Preservation of British Transport Relics*, p. 13. [17] ibid.
[18] Price, J H, and Pearson, F K, 1960, *British Tramcar Design, a report prepared on behalf of the Tramway Museum Society for the Consultative Committee on British Transport Relics.*
[19] Prior, Gareth, 2016, Picture in Time: Sheffield 342. Available at: http://www.britishtramsonline.co.uk/news/ (Accessed: 2 December 2021).
[20] Price, J H, 1946-1963, *Museum of British Transport Clapham,* (scrapbook of clippings), p. 144.
[21] *Minutes of a Meeting of the Tramways Sub-Committee of the Consultative Panel for the Preservation of British Transport Relics held at Clapham at 6:30pm on Tuesday, 24th October 1961,* pp. 2-3.
[22] Price, J H, 1970, *Consultative Panel for the Preservation of British Transport Relics*, pp. 8-9.
[23] Price, J H, 1964, *News from the Museums, in Modern Tramway and Light Railway Review, Volume 27, No. 319,* pp. 226-227.
[24] ibid., p. 227.
[25] Price, J H, 1970, *Consultative Panel for the Preservation of British Transport Relics*, p. 15.
[26] Price, J H, 1964, *News from the Museums, in Modern Tramway and Light Railway Review, Volume 27, No. 319,* pp. 226-227.
[27] Price, J H, 1967, *News from the Museums, in Modern Tramway and Light Railway Review,* Vol. 30, No. 352, p. 123.
[28] ibid., pp. 123-124.
[29] ibid., p. 124.
[30] Price, J H, 1970, *Consultative Panel for the Preservation of British Transport Relics*, p. 15.
[31] Prior, Gareth, 2016, *Picture in Time: Sheffield 342.* Available at: http://www.britishtramsonline.co.uk/news/ (Accessed: 2 December 2021).
[32] Price, J H, 1967, *News from the Museums, in Modern Tramway and Light Railway Review,* Vol. 30, No. 352, p. 124.
[33] *Minutes of a Meeting held at the Museum of British Transport, Triangle Place, Clapham, SW4 at 6:30pm on Monday 31st October 1966.*
[34] Price, J H, 1967, in *Modern Tramway and Light Railway Review*, Vol. 30, No. 356, p. 268.
[35] *Minutes of a Meeting held at the Museum of British Transport, Triangle Place, Clapham, SW4 at 6:30pm on Monday 31st October 1966.*
[36] ibid.
[37] ibid.
[38] *Modern Tramway and Light Railway Review,* 1967, Vol. 30, No. 360, p. 409.
[39] *Modern Tramway and Light Railway Review,* 1973, Vol. 36, No. 421, p. 11.
[40] Transport Act 1962, *British Historical Relics Scheme 1963, Schemes approved by the Minister of Transport in accordance with paragraph 1(5) of the Sixth Schedule to the Transport Act 1962. British Transport Historical Relics Scheme 1963, Transport Act 1926 Prepared by the British Railways Board. 2.-(a),* p. 1.
[41] *Transport Act 1962, British Transport Historic Relics Scheme 1963, 3.-(c), (e), 4.- (a),* p. 2.
[42] *Modern Tramway and Light Railway Review,* 1973, Vol. 36, No. 421, p. 11.
[43] *Modern Tramway and Light Railway Review,* 1969, Vol. 32, No. 373, p. 27.
[44] *Modern Tramway and Light Railway Review,* 1969, Vol. 32, No. 380, p. 284.
[45] Price, J H, 1972, *Museum News, in Modern Tramway and Light Railway Review,* Vol. 35, No. 415, p. 231.
[46] ibid., pp. 231-232.
[47] *Modern Tramway and Light Railway Review,* 1973, Vol. 36, No. 421, p. 11.
[48] ibid.
[49] *The Journal of the Tramway Museum Society*, 1973, Vol. 12, No. 64, p. 46.
[50] ibid., p. 47.
[51] ibid., p. 46-47.
[52] ibid., p. 47.
[53] *The Journal of the Tramway Museum Society,* 1973, Vol. 12, No. 65, p. 63.
[54] Bale, H, 2017, *Conversation with Peter Bird, Rolling Stock Engineer,* 28 September.
[55] *Modern Tramway and Light Railway Review,* 1973, Vol. 36, No. 421, p. 13.
[56] ibid.
[57] Sykes, R, 2021, email to Hannah Bale, 24 September.
[58] *The Journal of the Tramway Museum Society,* 1975, Vol. 14, No. 72, p. 31.
[59] Markham, J, 1976, letters to and from Anthony Bacon, September.
[60] *The Journal of the Tramway Museum Society,* 1976, Vol. 15, No. 78, p. 77.
[61] *The Journal of the Tramway Museum Society,* 1978, Vol. 17, No. 85, p. 53.
[62] Shawcross, J, 2014, *Collections Management Attitude Statement London County Council Tramcar Number 1,* LCC1/LIB/001, The National Tramway Museum, UK, p. 7.

[63] ibid.
[64] ibid.
[65] Bale, H, 2017, *Conversation with Peter Bird, Rolling Stock Engineer,* 27 October.
[66] Markham, J, 1976, letter to Anthony Bacon, 21 September.
[67] ibid.
[68] Ross, I, 1991, letters to and from Anthony Bacon, January.
[69] Ross, I, 2006, letter to Colin Heaton, 1 May.
[70] Ross, I, 2003, letter to Malcolm Wright, 27 November.
[71] LCCTT, 2006, letter to Colin Heaton.
[72] Ross, I, 2006, letter to Colin Heaton, 1 May.
[73] Dougill, I, 2006, letter to Ian Ross, 16 August.
[74] Ross, I, 2007, letter to Glynn Wilton, 23 April.
[75] Bale, H, 2017, *Conversation with Lawrence Dutton, Volunteer Coach Painter,* 26 May.
[76] Heeley, D J, 2012, *Tramcar Condition Assessment – London County Council No. 1*, 12/TCC/DJH/01, The National Tramway Museum, UK, p. 5.
[77] ibid.
[78] ibid.
[79] ibid., p. 6.
[80] ibid.
[81] ibid., p. 7.
[82] ibid., p. 5.
[83] ibid., p. 8.
[84] ibid., pp. 9-11.
[85] ibid., p. 9.
[86] ibid., p. 10.
[87] ibid., p. 11.
[88] ibid., p. 10-11.
[89] ibid., p. 8.
[90] ibid., p. 10.
[91] ibid., p. 5.

Chapter 9
[1] London County Council Tramways Trust, Trust Deed, 14 February 1988.
[2] Heeley, D J, 2012, *Tramcar Condition Assessment – London County Council No. 1,* 12/TCC/DJH/01, The National Tramway Museum, United Kingdom p. 5.
[3] Sturgess, N J, and Wilton, G, 2012, *LCC 1 Collections Management Attitude Statement*, 07/LIB/016, The National Tramway Museum, UK.
[4] Tramway Museum Society, 2012, 'Appendix 5 Curatorial Matters 18397'. *Minutes of the Board of Management 8th December 2012, Crich, Derbyshire* pp. 4-6.
[5] Shawcross, J, 2014, *Collections Management Attitude Statement London County Council Tramcar Number 1,* LCC1/LIB/001, The National Tramway Museum, UK.
[6] Stead, R, 2015, LCC1 – *The finer details*, in *The Tramway Museum Society Journal*, No. 229, pp. 31-33.
[7] Ross, I, 2014, *Bluebird – Off & Running!*, in *The Belvedere Echo*, No. 44, p. 1.
[8] Ross, I, 2015, *Bluebird In Bits*, in *The Belvedere Echo,* No. 45, p. 2.
[9] Ross, I, 2015, *Bluebird In Bits*, in *The Belvedere Echo*, No. 46, p. 2.
[10] Ross, I, 2018, *Bluebird Arises*, in *The Belvedere Echo,* No. 50, pp. 1-3.
[11] Ross, I, 2019, *Bluebird is going somewhere…,* in *The Belvedere Echo*, No. 52, p. 2.
[12] Ross, I, 2018, *Bluebird Arises*, in *The Belvedere Echo*, No. 50, pp. 1-3.
[13] Ross, I, 2018, *Nevertheless It Moves*, in *The Belvedere Echo*, No. 51, pp. 1-6.
[14] Ross, I, 2018, *Bluebird Arises*, in *The Belvedere Echo*, No. 50, pp. 1-3.
[15] ibid.
[16] Crabtree, M, 2017, *LCC1 – The Trucks [Part 1]*, in *The Tramway Museum Society Journal*, No. 237, pp. 19-21.
[17] Crabtree, M, 2017, *LCC1 – The Trucks [Part 3]*, in *The Tramway Museum Society Journal*, No. 239, pp.36-39.
[18] Crabtree, M, 2017/2018, *LCC1 – The Trucks [Part 4],* in *The Tramway Museum Society Journal*, No. 240, pp. 23-26.
[19] Crabtree, M, 2017, *LCC1 – The Trucks [Part 2],* in *The Tramway Museum Society Journal*, No. 238, pp. 25-27.
[20] Ross, I, 2017, *Bluebird: A Work In Progress…*, in *The Belvedere Echo*, No. 49, pp. 1-5.
[21] Crabtree, M, 2017, *LCC1 – The Trucks [Part 2]*, in *The Tramway Museum Society Journal,* No. 238, pp. 25-27.

[22] ibid.
[23] Ross, I, 2016, *Bluebird is Back*, in *The Belvedere Echo*, No. 48, p. 3-6.
[24] Crabtree, M, 2017, *LCC1 – The Trucks [Part 2]*, in The Tramway Museum Society Journal, No. 238, pp. 25-27.
[25] Crabtree, M, 2018, *London County Council Tramways 1 The Trucks – Part 5 – The End is in Sight!,* in *The Tramway Museum Society Journal*, No. 242, pp. 80-81.
[26] Ross, I, 2015, *Where Does that Wire Go*? in *The Tramway Museum Society Journal*, No. 230, pp. 14-15.
[27] Crabtree, M, 2017, *LCC1 – The Trucks [Part 2]*, in *The Tramway Museum Society Journal*, No. 238, pp. 25-27.

Appendix 2: Hidden Stories
[1] Brown43, 2019, *Savage Fitness and Self Defence*. Available at: https://www.tapatalk.com/groups/savagefitnessandselfdefense/rheo-blair-story-t2102.html (Accessed: 2 December 2021).
[2] Alston Garrard & Co. Ltd, 2019, *Liqufruta*. Available at: https://liqufruta.com/ (Accessed: 2 December2021).
[3] Bradstreet, Andrew, 2006, *Queen's Park History of the Spa.* Available at: http://www.mybrightonandhove.org.uk (Accessed: 24 August 2018).
[4] Haywards Pickles, 2021, *About us*. Available at: https://www.haywardspickles.co.uk/about-us/ (Accessed: 3 December 2021).
[5] Pool, Robert, 2014, *Original 'Camp Coffee' label*. Available at: https://www.bbc.co.uk/ahistoryoftheworld/objects/XG1CiGSCTzqb05nDwIhhjg (Accessed: 3 December 2021).
[6] Craft Metropolis, 2017, *Watneys – how to bring a bad beer back...but better.* Available at: https://www.craftmetropolis.co.uk/watneys-bring-bad-beer-back-better/ (Accessed: 2 December 2021).
[7] Cornell, Martyn, 2021, *The Hunting of the Stout.* Available at: https://zythophile.co.uk/2008/02/14/the-hunting-of-the-stout/ (Accessed: 2 December 2021).
[8] ibid.
[9] Grace's Guide, 2020, *Marconi's Wireless Telegraph Co.* Available at: https://www.gracesguide.co.uk/Marconi%27s_Wireless_Telegraph_Co (Accessed: 3 December 2021).
[10] Morleys in Brixton, *About Morleys*. Available at: https://www.morleysbrixton.co.uk/about-morleys/ (Accessed: 3 December 2021).
[11] Morleys in Brixton, *Store Info*. Available at: https://www.morleysbrixton.co.uk/store-info/ (Accessed: 3 December 2021).
[12] Swain, Alan, 2021, *Harringay Arena & Stadium,* Tottenham-Summerhill Road. Available at: https://tottenham-summerhillroad.com /harringay_arena_stadium.html (Accessed: 3 December 2021).
[13] GOV.UK, 2021*, London Assurance (The)*. Available at: https://find-and-update.company information.service.gov.uk/company/ ZC000054 (Accessed: 3 December 2021).
[14] Wikimedia , 2021, *Alliance & Leicester*. Available at: https://en.wikipedia.org/wiki/Alliance_%26_Leicester (Accessed: 3 December 2021).
[15] GOV.UK, 2021, *Wood's Bacon Factory Limited*. Available at: https://find-and-update.companyinformation.service.gov.uk/company/00347904/officers (Accessed: 2 December 2021).
[16] Ellis, Eric, *The 1950s*. Available at: http://www.mirfieldmemories.co.uk/1950s.htm (Accessed: 2 December 2021).
[17] Peach, B., *A Visit to Wood's Bacon Factory.* Available at: http://www.mirfieldmemories.co.uk/schools/mms/mms57mag04.htm#piggy_wood (Accessed: 2 December 2012).
[18] Secret Leeds, 2008, *Wallace Arnold.* Available at: https://www.secretleeds.com/viewtopic.php?t=1069 (Accessed 2 December 2021).
[19] Wikimedia, 2021, *Sixpence (British Coin)*. Available at: https://en.wikipedia.org/wiki/Sixpence_(British_coin) (Accessed 2 December 2021).
[20] Grace's Guide, 2017, *John Player and Sons*. Available at: https://www.gracesguide.co.uk/John_Player_and_Sons_(Player%27s) (Accessed: 23 August 2018).
[21] Wikimedia, 2020, *Stanford Robinson*. Available at: https://en.wikipedia.org/wiki/Stanford_Robinson (Accessed: 8 October 2021).
[22] London Fandom, *George Bryant Naish*. Available at: http://london.wikia.com/wiki/George_Bryant_Naish?oldid=60472 www.london.wikia.com (Accessed: 20 September 2018).
[23] ibid.
[24] Imperial War Museum, 2021, *British First World War Service Medals*. Available at: https://www.iwm.org.uk/history/first-world-war-service-medals (Accessed: 11 December 2021).
[25] Pointer, Ray, 2014, *Mutt and Jeff: The Original Animated Odd Couple*. Available at: https://www.traditionalanimation.com/2014/mutt-and-jeff-the-original-animated-odd-couple/ (Accessed: 8 December 2021).
[26] Imperial War Museum, 2021, *British First World War Service Medals.* Available at: https://www.iwm.org.uk/history/first-world-war-service-medals (Accessed: 8 October 2021).

Acknowledgements

We would like to thank the following for their willingness to share their advice, knowledge, memories, documents, photographs or other materials:

The late Paul Abell, Andrew Bailey, Peter Bird, John Buckle, Dr Paul Collins, Mike Crabtree, Arthur Dawson, Lawrence Dutton, Bill Fronczek (Pennsylvannia Trolley Museum), Michael Gilks, Dan Heeley, Frank Hicks (Illinois Railway Museum), John Horne, Peter Howard, Dave Jones, John Laker, Greg Marsden, David Packer, Alan Pearce, Ian Ross, Terry Russell, David Senior, the late John Shawcross, Mike Skeggs, Eric Smith, the late Jim Soper, Chris Thornburn, Peter Whiteley, Malcolm Wright, Derrick Yates and Ian Yearsley.

London Transport Museum
London Metropolitan Archives
London County Council Tramways Trust Collection
The National Tramway Museum Conservation Workshop
The TLRS Collection

In researching this book we have found the London Metropolitan Archives to be a valuable source of information. As home of the London County Council archive, the LMA looks after many photographs and posters relating to LCC tramways. A significant amount of this material is available to view online through the London Picture Archive. If you are interested in exploring this fascinating collection further, visit: https://www.londonpicturearchive.org.uk/

An equally valuable reference source has been the archive held by the London Transport Museum, which can be searched here: https://www.ltmuseum.co.uk/collections/

Author Biographies

Hannah Bale has been employed as Curatorial Assistant at the National Tramway Museum since 2014, assisting with the care of the Museum's diverse collections and research for the rolling programme of exhibitions. Having left school to work in the retail sector, she embarked on a change in career and has since gained a BA (Honours) History and MA Country House Studies from the Universities of Derby and Leicester respectively. Coming to the Museum as an intern in 2011 and later as a volunteer, she has also undertaken extensive voluntary work for the National Trust, Derby Museums and the Old House Museum, Bakewell. She is also employed at Haddon Hall.

Jim Dignan has been a volunteer at the National Tramway Museum since 2011, undertaking research and contributing to a variety of physical and online exhibitions. He has also been responsible for producing detailed profiles on each of the 82 tramcars that are featured on the museum's website. Before this he pursued a career in academia (Universities of Sheffield and Leeds), during the course of which he published numerous books, chapters and journal articles in the fields of penology, criminal justice and youth justice.

Robert Morris has worked as Librarian at the National Tramway Museum since 2016 and is responsible for the large collection of books, magazines, photographs and films in the Museum collection. He has contributed to the research and organisation of several exhibitions. Prior to this he worked for many years as a professional photographer, art director and picture editor. He has also worked as a Librarian for the National Trust Photographic Library and Greenpeace UK.

Ian Ross joined the Tramway Museum Society as a volunteer in 1967 becoming involved in many aspects of tramway engineering including tramcar restoration, overhaul, and maintenance. In 1982 and again in 2013 he was elected as a member of the Museum's Board of Management. He has also been a supporter of the London County Council Tramways Trust since 1970 becoming a Trustee in 1985 and Chairman in 2002. The Trust has provided the funds for the restoration of 'Bluebird' through donations from its supporters and from its trading subsidiary company. Ian has taken a leading part in the restoration of 'Bluebird' both in hands-on activities and in management of the project. Ian is a chartered electrical and mechanical engineer with a long career in the rail industry, retiring in 2009.

Richard Sykes With a lifelong interest in all forms of transport and with an overriding focus on tramways, Richard Sykes maintains an academic approach to the study of tramways both past and present. His researches continue to take him to tramway systems and museums worldwide to learn of, experience and record historic and current technological and operational aspects of tramways. Richard has for over 55 years physically contributed to the development of the National Tramway Museum and continues to actively participate in the preservation and restoration of tramcars within the Museum collection.

Lynn Wagstaff was educated at The Herbert Strutt Grammar School, Belper, and later, as a mature student, at the University of Derby where she gained a First Class Joint Subjects Degree in History and Theory of Design and Experience of Writing, followed by an MA in Narrative Writing. A regular volunteer at the Museum since 2010, from 2011 she has been part of the Library and Curatorial team, undertaking numerous tasks including writing and copy-editing text for publications and information boards. She also served for several years as Assistant Editor and Copy Editor of the Museum's Journal, and in early 2022 was appointed Journal Editor.

Laura Waters joined the National Tramway Museum as Collections Access Assistant in 2010, becoming Curator in April 2013. As Curator, she had overall responsibility for managing the Museum's physical and digital collections as well as its archives and exhibitions. An important part of her remit was to disseminate information relating to the Museum's collection to a wide variety of audiences across a range of media including physical and digital publications and displays. She also managed a curatorial team of employees and volunteers with wide-ranging knowledge and experience relating to various aspects of tramway history and operations. In 2020 she embarked on a fixed-term secondment with the Science Museum Group to help co-ordinate the relocation of over 300,000 artefacts from its London store to its National Collections Centre in Wiltshire. In 2021 she subsequently took up the permanent position of Group Collections Storage and Logistics Manager with the Science Museum Group.

General Index

Acton Works	114
Advisory Board of Engineers	32, 34
Advisory Committee on London Traffic	35
Anderson, Oliver	103, 107
Apprentice training	114
Ashfield, Lord	24, 25, 57
Baker, Alfred	52
Balham Station	115
Barrie, John	90
Bayswater Road tramway	28-9
Beaulieu	114
Beeching, Dr R	113
Bell, Alexander Graham	124
Belvedere Echo, The	118
Birkenhead	28
Birmingham Corporation Transport	52, 68-9
Blackfriars Bridge	34, 96
Blackpool Corporation Tramways	86
Bluebird	
Origin of name	10
'Blue Bird' car	10
Board of Trade	22, 26, 32, 33, 34, 43, 51, 66
President	32
Bright, John	32
British Electric Traction Co.	14
British Railways Board	113-5
British Transport Commission, The	6, 109, 111-2, 113
Consultative Panel	111, 113
Museum Board	113
Robertson, Sir Brian, Chairman	111
Tramway Sub-Committee	112
Bruce, Joshua Kidd	45, 55-6, 71
Buses	See Omnibuses, below
Bus operators	24, 26, 35-7, 39, 42, 47
Campbell, Sir Malcolm	10
Change Pit	30-1
Charles Roberts & Company	102
Charlton Works & Maintenance Depot	10, 20, 56, 75, 77, 79, 81, 82, 87, 92, 97, 99, 102, 110, 112, 140, 142, 145, 175 See also *Depots, Tram (includes Works)*, below
Chelsea Metropolitan Borough Council	20, 28
City & South London Railway Co.	
Coach	113
City of London	19-20, 21, 28, 45, 47
Collins, Stan	90-1, 96, 99
Compagnie Générale des Omnibus de Londres	36
Competition from motor buses	9, 14, 16
Competition from trolley buses	52-3
Conduit system (and equipment)	30-1, 55, 71, 89, 121, 128, 169, 170, 175, 176, 178-9, 188, 190
Conservation	115, 118-9, 123-8, 180
County of London	19-20
County of London Tramways Syndicate	42
Councils	
Chelsea Metropolitan Borough Council	20, 28
Croydon Borough Council	30

Greater London Council	114
Hackney Borough Council	30
Hertfordshire County Council	20
Ipswich Corporation	52
Kensington Metropolitan Borough Council	20, 28
Lambeth Borough Council	30
Leeds City Council	12
Lewisham Borough Council	30
London County Council	9, 10, 19, 45, 71, 134, 136, 140
Middlesex County Council	20
Stepney Borough Council	30
Walthamstow Municipal Borough Council	20, 27
Wandsworth Borough Council	30, 50
Westminster Borough Council	20, 28
Courtney, T	115
Crich	8, 44, 114-6, 118, 121-3
Crouch, J P	50
Croydon Corporation Tramways	20
Crystal Palace	115
Racing circuit	115
Station	115
Current collection (conduit)	26, 30-1, 71, 75
Current collection (overhead)	26, 30-1
Daimler buses	37, 40
Darwen Corporation Tramways	86
Dead-end terminals	32
Department of Education and Science	113
Depots, Tram (including Works)	
Leeds depots	
Chapeltown	103-4, 107
Kirkstall Road	102-3, 128
Swinegate	107-9
London depots	
Acton	25, 114
Brixton Hill	99
Depot fire	103
Camberwell	82, 89
Central Repair Depot	63, 77-8, 87, 92, 99
Charlton Works & Maintenance Depot	10, 77, 79, 87, 92, 97, 110
Fulwell Depot & Works	87
Holloway	90-1, 92, 94-6
Streatham	See *Telford Avenue Depot,* below
Telford Avenue Depot	90-2, 96
Derby	115
Dingle Underground Station	40
Disposal	113-4
Doncaster	11
Ealing Common	115
Edinburgh	70, 112
Edwards, E H	87
Electric Railway, Bus and Tram Journal, The	82, 87, 186
Electro-Mechanical Brake Company Limited	78
Elliott Hauliers Ltd	115-6
Elliott, Richard	6, 91, 103
English Electric Company Limited	64, 86, 113
Erith Council Tramways	20
'Evening Star' locomotive	114

Fares	11, 22, 26, 41-3, 45, 47, 55, 57, 72
Fell, Aubrey Llewellyn Coventry	45, 55
Feltham tramcars	16, 22-3, 58, 61-2, 68, 87, 91, 92, 98, 113-4, 115, 142
Feltham tramcars in Leeds	97, 102, 107, 142
Purchase from London Transport	102-3
Findlay, A B	102
Firth Brown Steels, Sheffield	103
'Frontagers'	34
Gateshead & District Tramways Co.	14-5
General Election 1931	22
Glasgow Corporation Transport	13, 86
Glasgow Museum of Transport	115
Global pandemic	44
Great Depression/Great Stock Market Crash	21-2
Grooved rail	26
Hackney Borough Council	30
Hall, Sir Benjamin	28
Health and Safety legislation	124
Hertfordshire County Council	20
Hilditch, Geoffrey	102, 107
HM Railway Inspectorate	125
Hopkins, C	12
Horse-drawn tramways	See *Tramways, horse-drawn,* below
House of Commons Select Committee	32
House of Commons Select Committee on Transport 1919	35
Houses of Parliament Joint Select Committee 1901	34, 43
Hume, G H (JP)	47
Hurcomb, Sir Cyril (Lord Hurcomb)	111
Hurst, Nelson and Company	46, 67-8, 74
Ilford Corporation Tramways	20
Ipswich Corporation Transport	52
Ireland, W E	6, 45, 51
Jackson, Mr (tram driver)	91, 92
Job Creation Programme	118
Kensington Metropolitan Borough	20, 28, 38
Kingsway Subway	2, 9, 18, 20, 30, 42, 66, 71, 82, 89, 96, 100
LCC No. 1	See *Index for LCC No. 1 (Bluebird)*
Leeds 301	
Alterations	103-4
Arrival in Leeds	99, 103
Bogies	103
Bow rope tubes	104
Braking system	103-4
Catching fire	103, 105, 107, 142
Departure to London	109-110
Electrical problems	103-4
Light Railway Transport League Tour, 1952	106-7
Mechanical problems	103-4
Operational problems	104
Paint scheme	103
Purchase from London Transport	99, 103
Repaint	103
Routes	104, 107, 108
Staff reaction to	103, 104, 107-8
Withdrawal from service	108

Leeds City Council
    Special Investigations Committee 108
Leeds City Transport 97, 102-3
    Transport Committee 97, 102, 107
    Leeds tramcar fleet, poor condition 102
    Running down of Leeds tram system 108-9
    Student conductors in Leeds 107-8
Leeds and District Transport News 104, 109
Lewisham Borough Council 30
Leyland Motors 40
Light Railway Transport League (LRTL) 111
    Museum Committee 112
    Tour 96-7, 106-7
Liverpool 36
Liverpool Corporation Tramway 12, 82, 86
Liverpool Edge Lane Works 82
Liverpool Overhead Railway 40
Local authorities
    Powers of veto 28, 30, 32, 33-4, 45
Locomotives
    'Evening Star' 114
London
    Geographical spread 19
    Population 19
    Size 19, 27
London Class HR/2 tramcar bogie 63, 64
London County Council 45, 71, 74
    Finance Committee 22
    Highways Committee 9, 21, 42, 45, 51, 55, 56, 63, 64, 65, 71, 72, 74-5, 78, 80, 87
    Housing policy 22
    Moderates Group 42, 43
    Municipal Reform Party 43
    Power to operate buses 20-1, 42, 47
    Progressive Group 40, 41-2, 43
    Social policy 11, 35, 41-2
    Tramways Department 10-1, 39, 45
    Transport policy 43
London County Council Tramways
    Absence of through-running arrangements 20, 34
    Closure 11
    Competition for new design of tramcar 50-1
    Deficit crisis 22
    Depots (including design and layout) 71-2
    Experimental tramcars
        LCC No. 1 .... See *Index for LCC No. 1 (Bluebird)*
        LCC 1852 .... 63-4, 67
        LCC 1853 .... 63-4, 67-8
    Fleet renewal 22, 45-6, 63-6
    General Order No. 709 90
    Highways Committee .... See *London County Council,* above
    Horse bus operations 36
    Manager 45-6, 65, 71, 74, 82, 87
    Modernisation programme 10, 18, 55-6, 63-6
    Modified tramcars
        LCC 1235 56
        LCC 1817 55-6
        Pullman cars 56

New tramcars	45-6, 63-4
Passenger numbers	47
Publicity	57
Pullmanisation programme	18, 56
Revenue (including shortfall)	43, 46-7
Traffic demand	45
Tramcar trailers	45-6
Tram fares	22, 42, 43, 45, 47
Tram-free vacuum (including doughnut-shaped hole)	36, 42
Workmen's tickets or fares	26, 42, 45
London County Council Tramways Trust (LCCTT)	6, 7, 118-9, 123
London Electric Railway Company	24
London General Omnibus Co. (LGOC)	23, 28, 36, 47
London Government Act, 1899	19
London and Home Counties Traffic Advisory Committee	24
London Motor Omnibus Company	40
London Passenger Transport Act 1933	35, 92
London Passenger Transport Board	11, 24, 41, 92
London Road Car Company	39
London Street Tramways Company	19, 42
London Traffic Act, 1924	35, 47
London Traffic Authority	35
London Traffic Board	43
London Tramways Company	19
London Tramways Syndicate	42
London Transport	
Courtney, T	115
London Transport Executive, The	97, 102, 111
London Transport Schedule	99
London Transport Supplies Meeting 1951	99
London United Tramways	16, 17, 20, 40, 42, 52, 57, 66
London Tramways Company	19
LPTB No. 1	See *Index for LCC No. 1 (Bluebird)*
LTE No. 1	See *Index for LCC No. 1 (Bluebird)*
MacDonald, Sir Ramsay	22, 24, 35
Makewell, Roy	156
Maley & Taunton	121
Manchester	32
Manchester Corporation Tramways	102
Mather, Alderman Bertrand	108
Matterface, Victor	6, 102-3, 107, 109
Metropolitan Electric Tramways	16, 17, 20, 40, 51, 57, 86
Experimental tramcars	
MET 318 'Bluebell'	16, 57-8, 66, 68, 78
MET 319/LUT 360 'Poppy'	57, 59, 66
MET 320	17, 59-60, 68
MET 330	58-9, 60, 68
MET 331	58-9, 61, 68
Metropolitan Police	40, 48
Metropolitan Railway	39
Metropolitan Vickers	74-5, 78
Middlesex County Council	20
Milan Tramway (Azienda Trasporti Milanesi [ATM])	14
Minister of Transport	35
Ministry of Transport	22, 51, 66
Model tramcar	10, 78-9, 91, 112-3
Moderate Party	See *London County Council,* above

Modern Tramway	97, 113
Morgan, J P	40
Morrison, H, Lord	24
Motor buses	See *Omnibuses,* below
Municipal Reform Party	See *London County Council,* above
Municipal obstructionism	28-32
Municipal tramways	11, 21, 25
Municipal Tramways Association	51, 57
Municipal Tramways & Transport Association	25
Municipalism (doctrine of)	11, 42

Museums
- Beamish ... 114
- British Transport Museum, Clapham ... 6, 113-6, 118, 121
  - Consultative Panel ... 6, 111, 113
  - Exhibition ... 112, 113
    - Of Small Objects ... 112
    - Of Large Relics ... 112
    - 'Tramway Avenue' ... 112
  - Preservation of British transport relics ... 111
  - Tramcars considered for disposal ... 113-4
- Tramway Sub-Committee ... 112-3
  - Transport history ... 113
- East Anglia Transport Museum at Carlton Colville ... 163
- Glasgow Museum of Transport ... 115
- London transport museum proposal ... 114-5
- National Tramway Museum, Crich ... 6-8, 123, 129, 143
  - Century of Trams display ... 127
  - Great Exhibition Hall ... 126-7
  - Restoration/Conservation Workshop ... 129
  - Tramcar Conservation Committee ... 119
- Railway Museum, York ... 111, 112

National Government	22, 24
National Tramway Museum	See *Muscums,* above
North Metropolitan Tramways Company	19
Norwood News	114-5
Office of Rail and Road	125
Oldham Corporation Transport	52
Omnibus Magazine	87

Omnibuses
- AEC Regent chassis ... 49, 102
- B-type ... 37-8, 40, 47
- K-type ... 47
- NS-type ... 47-9
- ST-type ... 49
- Competition with tramcars ... 46-9
- Diesel-engined ... 44, 102
- Improvements in design & technology ... 47-9
- Increased numbers in London ... 35-7
- Leyland Titan ... 49
- Petrol-engined ... 37
- Popularity ... 37, 44
- Services in London ... 36-7
- Tilling Stevens bus chassis ... 115

Overhead wiring	26, 30
Packer, David	115
Paris Metro	40-1
Parliamentary Bills & Acts	

Acts	**19, 23-4, 26-7, 28, 32, 33, 35, 45, 47, 92, 113, 114**
Bills	**22, 24, 28, 32, 33, 35**
Penhall Road, Charlton	See *Depots, tram,* above
Pick, F	**24-5**
Pickfords Removals	**109**
'Pirate' buses	**24, 35**
Plough	**30-1, 65, 68, 75, 77-8, 99, 103, 118, 121, 128**
Preservation of Relics and Records, The	**111**
Presidents' Conference Committee (PCC)	**14**
Price, J H	**6, 10, 82, 112, 113, 114, 123**
Progressive Group	See *London County Council,* above
Public transport 'free-for-all'	**36-9**
Pullman Review, The	**82**
Pullmanisation programme	See *London County Council Tramways,* above
Rafferty, Alderman	**109**
Railways	See also *Underground railways,* below
Nationalisation	**111**
Suburban rail travel	**46**
Regulatory framework	**26-7**
Failure to reform	**32-6**
Riddles, R A	**111**
Robertson, Sir Brian	**111**
Robinson, P J	**82-3**
Rolling stock	**102, 111**
Royal Commission on London Traffic (1905)	**30, 34**
Royal Commission on Means of Locomotion and Transport in London	**45**
Royal Commission on Transport (1931)	**24-5, 72, 87**
Final Report of	**24-5**
Scholes, John	**111-2, 114**
Sheffield	**112**
Sheffield Corporation Tramways	**12, 69**
Short Brothers	**68**
Sinclair, G F	**6, 45, 82, 91**
Skeat, W O	**111**
Smith, Eric	**103, 107**
Smith, Tom	**103, 107**
South Metropolitan Electric Tramways & Lighting Co.	**20**
South Shields Tramways	**86**
Southampton Corporation Tramways	**102**
Southern Counties Touring Society	**2, 97**
Spencer, C J	**51, 57, 87**
Speyer, Edgar	**40-1, 43**
Standing Committee on Rolling Stock	**51**
Stepney Borough Council	**30**
Stephenson Locomotive Society	**111**
Steward, Tom	**103-4**
Stock Market Crash	See *Great Depression,* above
Sunderland Corporation Tramways	**86**
Thames, River	**19, 30, 42**
'The Combine'	See *UERL (Underground Group),* below
Thomas, T E	**45, 82, 87**
Thomas Tilling	**40**
Thornburn, Chris	**107-8**
Tickets (Workmen's)	See *Workmen's fares/tickets,* below
Tilling-Stevens bus chassis	See *Omnibuses,* above
Track density	**32**

Train, G F		**28-9**
Tramcars		
	Birmingham 842	**68-9**
	Birmingham experimental designs	**68-9**
	Blackpool 1	**114**
	Blackpool electric locomotive 717	**180**
	Bombay tramcars	**86**
	Cabin cars, Liverpool	See *Liverpool Cabin (or Robinson) cars,* below
	Central entrance	**72**
	Chesterfield 8	**114**
	Developments in tramcar bodywork construction technology	**66-70**
	Douglas Head Marine Drive 1	**114**
	Douglas 14	**114**
	E/1 car 1025	**113**
	E/1 car 1235	**56**
	Front exit, rear entrance	**15, 57, 72**
	Edinburgh 35	**127**
	Edinburgh experimental tramcar 180	**70**
	Glasgow 1392	**114-5**
	Glasgow Kilmarnock Bogies	**13**
	Grimsby and Immingham 14	**114**
	Grimsby and Immingham 26	**114**
	Horse-drawn	**26-8, 36, 42**
	Leeds 164	**108**
	Leeds 301	*See separate entry, above*
	Leeds Middleton Bogies	**12, 86**
	Liverpool 869	**143**
	Liverpool Liners Class	**13, 143**
	Liverpool Marks cars	**82**
	Liverpool Cabin (or Robinson) cars	**82-4, 86, 143**
	Llandudno 6	**114**
	London	
	Class A	**65, 88**
	Class B	**88**
	Class C	**63, 88**
	Class D	**65**
	Class E	**78, 86**
	Class E/1	**45-6, 49, 55-6, 63-4, 77-8, 84, 88, 113**
	Class E/3	**64-6, 72, 78, 88**
	Class HR/1	**63**
	Class HR/2	**63, 64, 65, 66, 72, 75, 78, 80, 88**
	Class ME/3	**84**
	Feltham	See *Feltham tramcars,* above
	London County Council No 1	See **Index for LCC No. 1 (Bluebird)**
	London County Council 106 (Snowbroom 022)	**114**
	London Transport 1852	**67-8**
	London Transport 1853	**65, 68**
	London Transport HR2	**66**
	London United Tramways 159	**119**
	LPTB 1622	**147**
	LTE 2144	**103**
	LTE 2162	**103**
	MET 331	**130**
	MET 355	**114, 115**
	Milan ATM Class 1500	**14**
	Newcastle 102	**114**
	New York 674	**151**

North Metropolitan horse tram 184	130
PCC streetcars	14-5
Robinson cars	See *Liverpool Cabin (or Robinson) cars,* above
Sheffield 342	112, 114
Sheffield experimental tramcar 370	69
Sheffield 501	102
Sheffield 510	126
Single entrance and exit	72
Snowbroom 022	See *London County Council 106,* above
West Ham 290	113
Tramway and Light Railway Society	112
Tramway and Railway World, The	71, 82, 86, 87, 186-90
Tramway Museum Society (TMS)	6, 112, 118, 123
Board of Management	7, 118, 123-4
Curator	123
Mission Statement	125
Tramcar Conservation Committee	123
Tramway Operators	
Challenges facing	9-10, 19-21, 26
Competitive disadvantages	23, 26-7, 36-9, 44
Local authority	11-3, 18, 26-7
London	23, 26, 28-31, 41-4, 45-6, 47, 49-52, 55-7, 63-6, 67
Municipal opposition	28-31
Overseas	13-4
Private/company-owned	14-7, 27, 42, 47, 52, 57-62, 66-7
Regulatory constraints	26, 32-6
Responses to challenges	10
Technical challenges	26
Veto powers of adjoining property owners or occupiers	27
Tramways *Individual tramway operators are indexed separately*	
Closures	10, 20, 25, 28, 44, 52, 97, 99, 103, 108-9
Electric traction	9, 26-8, 36, 42
Horse-drawn	26-8, 36, 42
Elsewhere	28, 36
In London	28-9, 42
Passenger numbers	36
London tramways	
Proposals for extension	34, 43, 52, 65
Modernisation	10, 12, 18, 50, 55-7, 63
Replacement by motor buses or trolleybuses	10, 44, 52-3, 92, 96, 109
Step rail	28
Tramways Act 1870	26, 28, 33, 45
Tramways and Light Railways Association	25, 51, 57
Tramways Bill 1870	32
Tramways, Trolley Vehicles and Motor Omnibuses Compared (LCC Report)	49-50
Tramways, Light Railways and Transport Association	82, 87
1932 Conference	87
Transport Act 1962	113, 114
Transport and General Workers' Union	25
Transport Trust	115
Trolleybuses	10, 14, 25, 44, 52-3, 62, 87, 92, 96, 102, 112, 115
Competition with tramcars	52
Turckheim, A de	87
UERL (Underground Electric Railways Group)	21, 23, 24, 43, 47, 57
Underground railways	21, 23, 26, 33, 34, 35, 39-41, 42-3, 57, 66
Competition with tramcars	55
Union Construction Co.	16, 57-8, 61

Vanguard Company	**39**
Walthamstow Corporation Tramways	**20**
Walthamstow, Municipal Borough	**27**
Wandsworth Borough Council	**30**
West Ham Corporation Tramways	**20**
Westminster Borough Council	**20, 28**
Westminster Bridge	**28**
Wood, T McKinnon	**43**
Workmen's fares/tickets	**26, 43-4, 45**
Yerkes, C T	**39-40**
Yorkshire Evening News	**109**

Index for LCC No. 1 (Bluebird)

Access handrail 103, 128
Adverts (including advert strip) 120, 156-7
Air brakes 75, 82, 91, 92, 104, 107
    Cylinders 75
Air equipment 75, 104, 118
Air brake pipe 103
Air operated doors 72, 82
Air operated platform steps 103, 128
Air operated windscreen wipers 103, 128
Alterations 103, 104, 125, 133
Artefacts found during restoration 191-8
As 'Bluebird' 6, 7, 8, 9, 10-1, 18, 21, 22, 44, 87-92, 99, 119
As LCC No. 1 .... 9, 10, 12, 14, 18, 21-5, 45, 75, 80, 82-6, 87-92, 102-3, 112-22, 123-80
        Photographs 2, 9, 31, 36, 41, 44, 77, 79, 81, 87, 90, 91, 93, 94, 95, 181
As Leeds 301   See **General Index**
As LPTB No. 1 .... 92, 96-7
As 'the experimental tramcar' 45, 71-82, 87, 92, 129, 133, 140
As 'the new tramcar' 87, 89, 92
As LTE No. 1 .... 97, 98-101
Asbestos 125, 130, 132-3
At Clapham 110, 112-6
Attitude Statement 124-5
Axles 74, 78, 80, 161-2, 172

Blanking panel 103, 121
'Bluebird' 44, 71, 136
        Origin of name 10
Body/Bodywork ... 76, 82, 84-5, 110, 119-20, 124, 126, 130, 133, 134-5, 143, 145, 151, 153-4, 157, 158,
.... 162, 165-6, 170, 180, 188, 190, 193
Body fittings 118
Bogies 74, 77, 80, 103, 110, 114, 145, 148, 157-60, 162, 165, 166, 169, 170, 172
Bolsters 74, 75, 80, 188
Bow collector 103, 115, 118, 128, 151, 179, 192
Brake systems
Brake chains 128
Brecknell, Willis & Co. Ltd. 118

Camira Fabrics 147
Canvas 120, 151
Carlines (metal roof joints) 121, 188
Ceiling (including ceiling panels) 75, 113, 121-2, 132-3, 139, 142, 155, 175, 179, 186-8, 196
CMS Cepcor 163
Chrome fittings 121, 140, 154, 155, 180
Compressor 128, 167, 169-72, 176, 179, 180, 190, 193
Condition Assessment, 2012 119-121
Conduit equipment 121, 128, 169, 170, 175, 176, 178-9, 188, 190
Conduit power supply electric equipment 169, 170, 175, 178, 179, 180, 190
Conservation 115, 118-9, 123-8
Conservation Workshop 7, 124-30, 135, 141, 142, 144, 146, 147, 148, 149, 151, 154, 157, 158, 160,
.... 170, 174, 179, 180, 191, 197
Construction 74-80, 135, 180
        Cost (including budget) 74-5
        Methods 65, 71, 74-8, 80, 139-40, 142, 143, 176, 186, 188
        Techniques 71, 74-8, 80, 82, 126, 132, 149, 188
        Timescale 74-5, 180
Control equipment 78, 121, 189
Controllers 75, 104, 118, 121, 122, 172-3, 176-9, 189
Corrosion 119-22, 124, 142, 145, 167, 169
Cove panels 133, 186

Dashes 120, 157

Design **71-4, 78, 87, 107, 118, 119, 124, 125, 126, 133, 140, 142, 143, 147, 148, 149, 151, 156, 180, 186, 188**
Destination and service number blinds 96, 103, 128, 155-6
Destination display boards 82
Deterioration 119-21, 128, 169
Dimensions 71, 78, 159, 186
Disposal 6, 99, 103, 114
Door motors 167, 169
Doors 75, 82, 103, 121, 126, 128, 147, 167, 180, 189, 190
Dorlec Limited 160, 161, 162
Driver's cab 82, 132, 167, 176, 186, 187
    Cab doors 121, 132, 147
    Door handles 132
    Hinged driver's seats 75, 142, 150

Electric bells 82, 179, 189
Electric buzzers 179
Electric gong 75, 189
Electrical equipment 121, 128, 180, 189
Electrical fires 103, 105, 107, 142, 176, 177, 191
Electrical wiring 105, 132
Electro-Mechanical Brake Company (EMB) 78, 120, 157
Electrolytic corrosion 119
Energy consumption 71
Equipment **65, 71, 74, 75, 78, 87, 104, 118, 121, 125, 128, 129, 132, 172-4, 179, 180, 189, 190**

Fares 92
Floor 71, 74, 75, 121, 143, 154-5, 156
    Floor covering 121, 122, 135, 155, 156
Folding step 82, 103, 121
    Folding step mechanism 150-1, 171, 172
Fuses/Fuse board 118, 121, 122, 175, 179, 187

Garmendale Engineering Co. 144, 145-6, 153
G D Peters & Co. Ltd 80
Grimsby and Cleethorpes Model Engineering Society 149
Grinsty Rail Limited 156

Handbrake linkage 147
Headstock 74-5, 188
Heating/Heaters 80, 82, 90, 133, 146, 176, 179, 187

Indicator boxes 82, 90, 186
Influence on tramcar design 82-4, 86
Interior panels/lining panels 133, 139, 191
Interlock circuit breaker 118, 178
John Holdsworth & Co. Ltd 147

Kervick, Trevor 157, 158

Launch 11, 18, 82, 89
LCC No. 1   See *As LCC No. 1, above*
Leeds 301   See **General Index**
Lifeguards 103, 115, 150
Lighting and light fixtures 71, 80, 90, 113, 126, 132, 179, 186, 187-8
Livery 10-1, 78-9, 82, 91, 92, 114, 115, 126, 131,134, 136, 137, 156-7, 186
    Ivory streamlining 10, 82, 156, 186
    London County Council crest 134, 136, 145, 157
Lower deck frame 74

Main frame 75, 120-1
Material coatings 119
Metal components (including aluminium, steel, etc.) .. **71, 74-6, 78, 82, 86, 92, 103, 113, 119, 120, 121, 122, 126, 128, 130, 133, 134, 135, 139, 140, 143, 145, 146, 147, 150, 154, 159, 160, 161, 163, 167, 169, 172, 178, 180, 186-90, 192**

Metropolitan Stage Carriage number	78
Metropolitan Vickers	74, 75, 78, 172
Model tramcar	10, 78-9, 91
Modifications	118, 119, 125, 139-40, 142, 164, 170, 178-9
Moquette	118-9, 147, 154, 156, 186, 187
Motorman's seats	75, 128, 142
Motorman's valve	104, 121, 170, 172, 190
Motors	74, 75, 77, 78, 118, 157, 160, 161-3, 165, 172, 179-80, 188, 189, 190
Number	78, 87-9, 92
Number blind boxes	92, 103, 118, 128, 140, 186
Overhaul	92, 97, 102, 103, 114, 118, 145, 159, 160, 162
Paint (including Paintwork)	78, 92, 103, 120-1, 134, 136-7, 145, 151, 153, 155-8, 160, 164, 186
Paint finishes	78, 120, 137, 140, 157, 160, 186
Performance	71, 123
Pillars	119-20, 140, 187
Plank/trolley bar	118
Plough gear (including plough assembly)	75, 77-8, 99, 103, 121, 128, 147, 157, 188
Plough carrier	77-8, 180
Plough-trolley changeover switch	118
Platform	74, 75, 78, 103, 104, 108, 120, 138, 139, 147, 167, 169, 180
Doors	75, 103, 126, 128, 147, 150, 191
Steps	75, 103, 126, 128, 148
Police lights	131, 132
Preservation	109, 111-22, 123, 124, 154, 157, 198
Press reviews	9, 80, 87
Repairs	92, 104, 121, 145, 176, 177
Restoration	103, 114, 118-9, 123-181
Fund-raising	114, 118-9, 123, 124
Labour resources	124, 128-9
Mobile crane	134
Project team	125, 127-8, 130, 143, 146, 147, 149, 156, 157
Specification	125-6, 147, 150, 151, 161, 178, 180
Storage space	130, 157, 172
Timespan	118-9, 127-8
Effect of Covid pandemic on progress of restoration	180
Restoration process	129-81
Accommodation bogies	134
Artefacts and curios found during deconstruction process	191-8
Asbestos removal	125, 130, 132, 133, 172, 176-7
Recanvasing of roof	151-3
Repainting	151, 157
Treatment of components, fittings and original materials	132, 143-51
Long lead-time projects	128
Missing items	122, 128, 148, 150
Move to Conservation Workshop	124, 126-7
Post-restoration commissioning programme	181-4
Recording process	129-30, 161, 168, 175
Restoration Programme	127-8
Restoration Progress	130-80
Work on air-supply system	167-72
Analysis	168-9
Deconstruction	167-8
Quick release valve	172
Reassembly	170-2
Treatment	169-70
Work on bodywork & associated mechanical work	130-58
Analysis	136-43
Deconstruction	130-5
Reassembly	151-7
Treatment	143-51
Work on electrical supply system and associated mechanical equipment	172-80

Analysis	175-6
Deconstruction	172-6
Reassembly	179-80
Treatment	176-9
Work on trucks and associated electrical work	157-66
Analysis	158-60
Axles	161-2, 172, 177
Brakes	120, 158-60, 169-71, 172, 179-80
Deconstruction	157-8
Motors & wheelsets	160-5
Reassembly	164-6
Treatment	160-4
Rexine tram facings	155, 186, 187
Roof	74-5, 120, 121, 128, 130, 133-5, 143, 151-4, 187, 188
Saloon, Upper and Lower	133, 137, 138, 140, 143, 147, 155, 175, 179, 186-9
Screws	133-4, 140
Sealants	119, 142
Seat frames	133
Seating (including capacity)	71, 72, 75, 80, 90, 119, 126, 130, 132, 140, 142, 147, 156, 186, 187
Driver's seats See *Driver's cab*, above	
Services	
London Route 10	96
London Route 16	95, 96
London Route 16 EX	95
London Route 18	96
London Route 18 EX	95, 96
London Route 22	96
London Route 33	89, 90
London Route 35	89, 90, 91
London Route 35A	89, 92
Subway	89, 96
Side panels	74, 92, 103, 128, 142, 143, 144, 145, 146, 154
Side plates	74, 75
Skirting boards	133, 176
Staircases	85, 121, 139, 140, 142, 175, 176, 180, 186, 187, 189
Switchgear	118
Tension springs	118
Testing and commissioning	180-3
Traction wiring	118, 179, 180
Trolley bases and booms	115, 118, 151, 179
Trolley wheels	151, 179
Truck mechanisms	157-66
Trucks	74, 77, 78, 80, 102, 103, 120, 128, 137, 158, 159, 161, 162, 164, 186, 189, 190
Experimental	77, 78, 80
Tube heaters	133, 146-7
Underframe	78, 80, 120, 130, 147, 175, 178, 188
Upholstery	118, 119, 186
Used ticket boxes	138
Contents	138
Ventilation/ventilators	65, 71, 90, 126, 139, 147, 187, 188, 191
Vestibules	82
Weight	71, 78, 115, 188
Wheels/wheelsets	74, 82, 84, 92, 103, 157, 159, 160, 161, 162
Windows (including sills)	74, 90, 92, 103, 120, 121, 128, 132, 138, 140, 142, 145, 148-50, 155, 187, 188, 189
Windscreens	103, 126, 143, 146, 157
Windscreen wipers	103, 126
Wood (including plywood etc.)	103, 121, 141, 143, 150-1, 155, 186
Woodwork	121, 150, 186